Sustainable Packaging Symposium (SPS 2019)

Santa Barbara, California, USA
16 - 18 December 2019

ISBN: 978-1-7138-0733-9

Sustainable Packaging Symposium (SPS 2019)

Santa Barbara, California, USA
16 - 18 December 2019

AIChE
The Global Home of Chemical Engineers

Printed from e-media with permission by:

Curran Associates, Inc.
57 Morehouse Lane
Red Hook, NY 12571

Some format issues inherent in the e-media version may also appear in this print version.

Copyright© (2019) by AIChE
All rights reserved.

Printed with permission by Curran Associates, Inc. (2020)

For permission requests, please contact AIChE
at the address below.

AIChE
120 Wall Street, FL 23
New York, NY 10005-4020

Phone: (800) 242-4363
Fax: (203) 775-5177

www.aiche.org

Additional copies of this publication are available from:

Curran Associates, Inc.
57 Morehouse Lane
Red Hook, NY 12571 USA
Phone: 845-758-0400
Fax: 845-758-2633
Email: curran@proceedings.com
Web: www.proceedings.com

TABLE OF CONTENTS

PLENARY SESSION

TECHNICAL SESSION 3: INNOVATIVE TECHNOLOGIES

Bouncing Back? Single Use Plastic Bans and the Potential for Rebound Effects

Conference: Sustainable Packaging Symposium (SPS)

Year: 2019

Proceeding: Sustainable Packaging Symposium 2019

Group: Abstract Submissions

Session: Technical Session 1 Packaging Sustainability Today

Time: Monday, December 16, 2019 - 11:45am-12:10pm

Authors:

Taylor, C., EarthShift Global
Laurin, L., EarthShift Global

The high visibility of plastics waste in vulnerable environments, especially marine and riparian areas, is leading to an expansion of bans on plastic shopping bags and single use plastics. But are these efforts actually moving us in the right direction? Evidence suggests not, via two factors that contribute to unintended consequences: leakage and the rebound effect. Providing heavier weight, longer lived plastic shopping bags, for example, appears to be simply increasing the amount of plastic waste, as the bags are treated the same as their precursors, and other rebound mechanisms abound. This presentation will discuss current and emerging evidence of both leakage and rebound in the wake of restrictions on plastics, and the mechanisms driving them. These have implications for both assessment and projections, and we will discuss how we might approach accounting for it, and how it might be limited or mitigated.

The US PET Bottle Market: A Mismatch between Product Circularity, Material and Environmental Efficiency

Conference: Sustainable Packaging Symposium (SPS)

Year: 2019

Proceeding: Sustainable Packaging Symposium 2019

Group: Abstract Submissions

Session: Technical Session 1 Packaging Sustainability Today

Time: Monday, December 16, 2019 - 12:35pm-1:00pm

Authors:

Lonca, G., CIRAIG

The Circular Economy (CE) movement is inspiring new governmental policies along with company strategies. This led to the emergence of a plethora of indicators to quantify the â €œcircularityâ€ of individual companies or products. Behind this approach at micro scale to quantifying circularity, the CE supports mainly two implicit assumptions we tested with a large-scale case study on the circularity of PET in the US market within the bottle market and across all sectors. First, closing the material loops at the micro scale within the same product is expected to lead to improvements in material efficiency for the economy as a whole. Second, maximizing material circularity contributes to mitigate environmental impacts. Material Circularity Indicator (MCI) reveals that closing the material loops at the product level increases product circularity and lead to higher material efficiency in the US plastic bottle market but not in the US PET market across all sectors using PET. Life Cycle Analysis (LCA) results reveal that rPET in bottle manufacturing seems environmentally beneficial from an attributional standpoint but not from a consequential, market-wide viewpoint. By adopting a broader approach than product level circularity, we contribute:

- to invalidate the assertion that improving one product improves products average and,
- to break the implicit link between environmental performance and material efficiency supported by the CE movement.

We further discuss methodological aspects for CE assessment.

Top 5 Pillars of Sustainable Paper-Based Barrier Packaging

Conference: Sustainable Packaging Symposium (SPS)

Year: 2019

Proceeding: Sustainable Packaging Symposium 2019

Group: Abstract Submissions

Session: Technical Session 1 Packaging Sustainability Today

Time: Monday, December 16, 2019 - 12:10pm-12:35pm

Authors:

Gasparik, T., Smart Planet Technologies

As global demand for plastic reduction, landfill waste diversion and increased recycling rates continues to grow, itâ€™s important that any innovative material advancement for sustainable paper barrier packaging address the following five areas:

Performance â€“ when it comes to a paper cup or foodservice package, it canâ€™t leak, should run on conventional high-speed converting equipment and needs to withstand the rigors of its intended use.

Environment â€“ the first step to getting paper cups and foodservice packaging recycled is to make them valuable and easy for recyclers to process into high-grade recycled paper products using conventional paper recycling equipment. (Less plastic, recycled content, etc. is also a priority.)

Economics â€“ nobody wants to pay more to be â€˜greenâ€™ but municipal/state/EPR mandates, consumer pressure and increased media coverage are moving companies in this direction.

Commercial Availability â€“ Alternative solutions need to be commercially available in order to be considered as a sustainable solution. Being asked by your packaging company to wait for a future product to come online is a stall tactic used by those who donâ€™t want anything to change, regardless of how better, faster, cheaper the new solution may be. Donâ€™t let the â€˜perfectâ€™ solution stand in the way of making â€˜greatâ€™ progress. We need incremental change in order to solve our waste problem. There is no silver (or should I say â€˜greenâ€™)

bullet.

Safety â€“ paper-based barrier products made using PFOA, BPA, BPS and heavy metals should not be used due to health concerns per recent studies.

sustainable development goals

a brief introduction

Sustainable Packaging Symposium
16 December 2019

Minal Mistry | Oregon Department of Environmental Quality

Source: UNDP

Millennial Development Goals (2000)

Source: UNDP

progress - India

Source: UNDP

progress - Bangladesh

MILLENNIUM DEVELOPMENT GOALS	INDICATORS	CURRENT STATUS	TARGET 2015
Eradicate extreme poverty	Proportion of people below national poverty line (%)	31.5 (2010)	29.0
		ON TRACK	
Universal primary education	Proportion of pupils starting grade 1 who reach grade 5 (%)	79.5 (2011)	100.0
		NEEDS ATTENTION	
Promote gender equality	Ratio of girls to boys in primary education	1.03 (2011)	1.0
		GOAL MET	
Reduce child mortality	Infant (0-1 year) Mortality Rate (per 1,000 live births)	35	31
		ON TRACK	
Improve maternal health	Maternal Mortality Ratio (per 100,000 live births)	194 (2010)	143
		ON TRACK	

Source: UNDP

Sporadic progress - overall

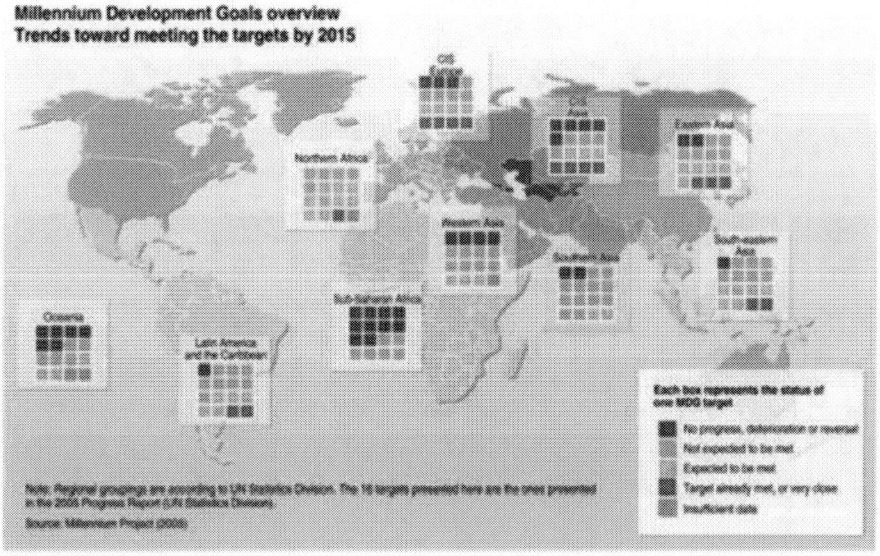

Source: UNDP

sounding the alarms

climate change came to the fore

Welcome to the Anthropocene

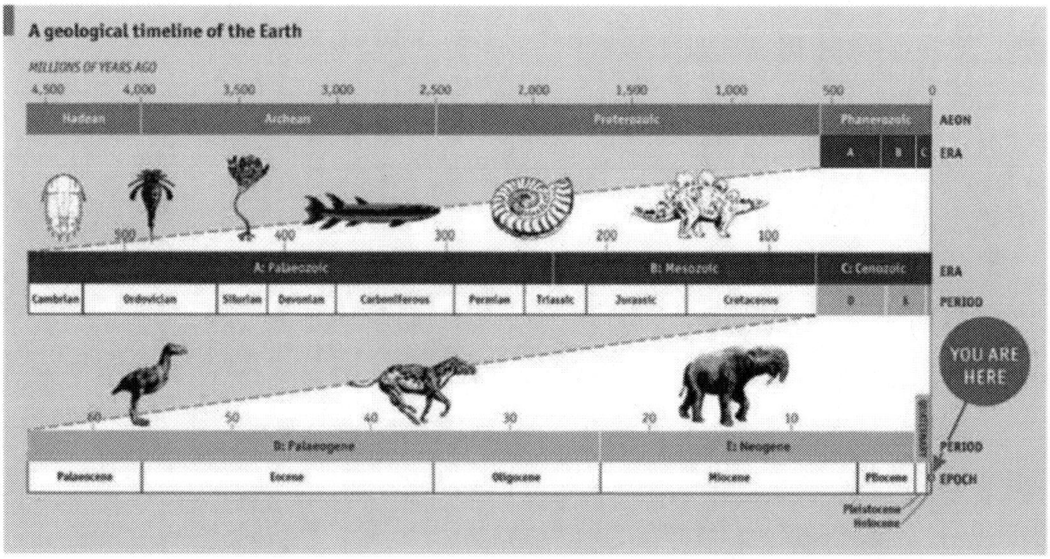

Source: The Economist

What changed since 2000?

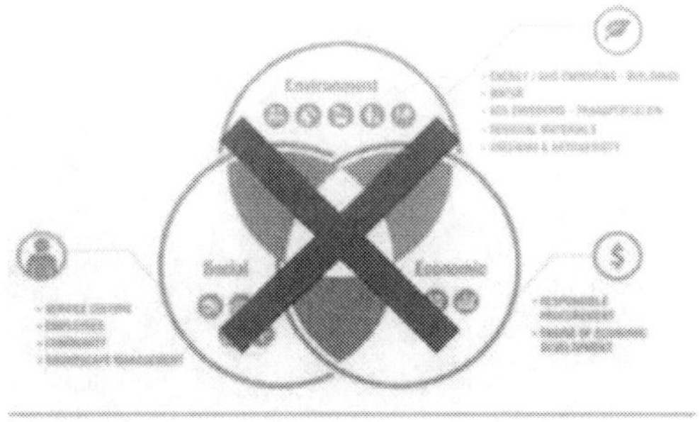

limits of a living planet

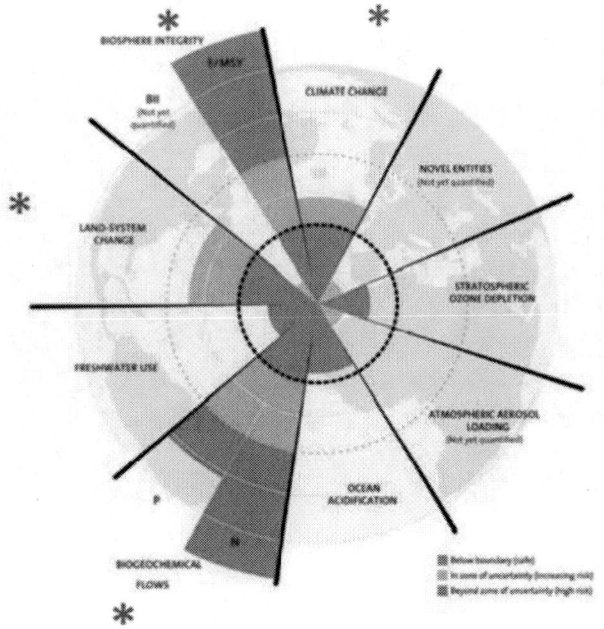

Source: https://www.stockholmresilience.org

Root cause: the economic growth imperative

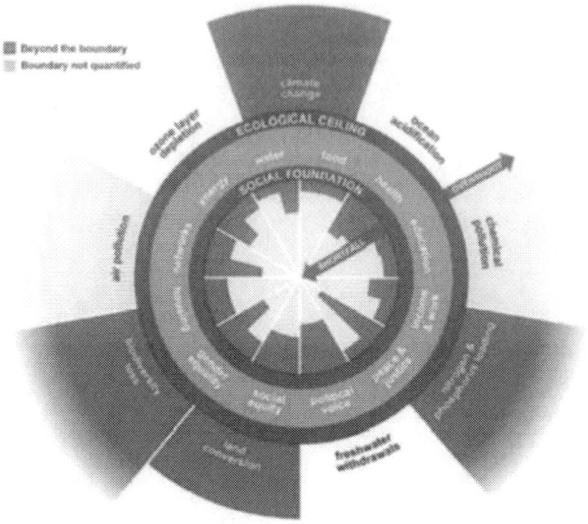

Source: Kate Raworth, Donut Economics

Real chasm between climate science and politics

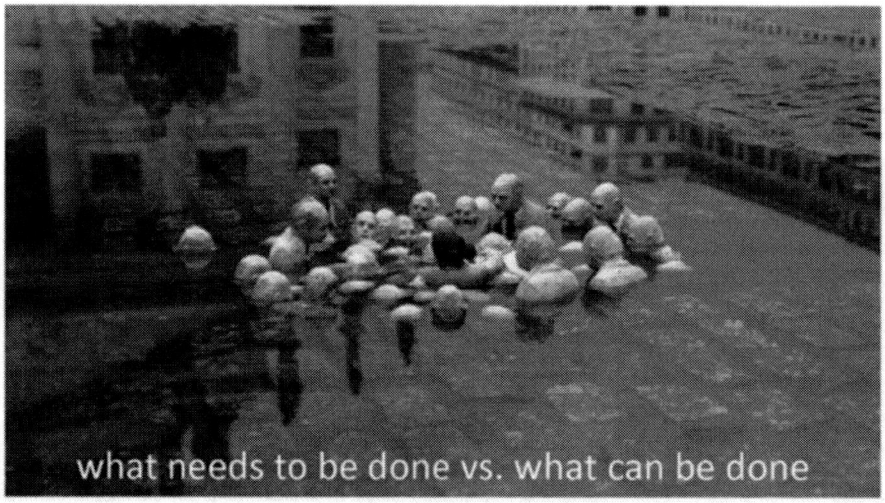

Source: PRI.org

Extreme poverty and human rights

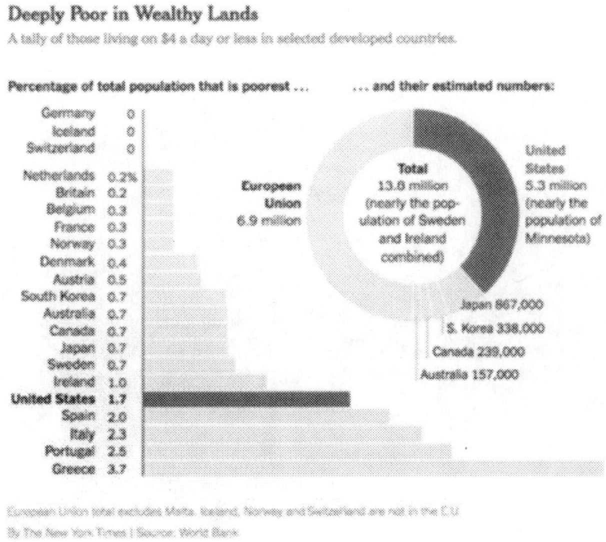

Need for *universal development* recognized

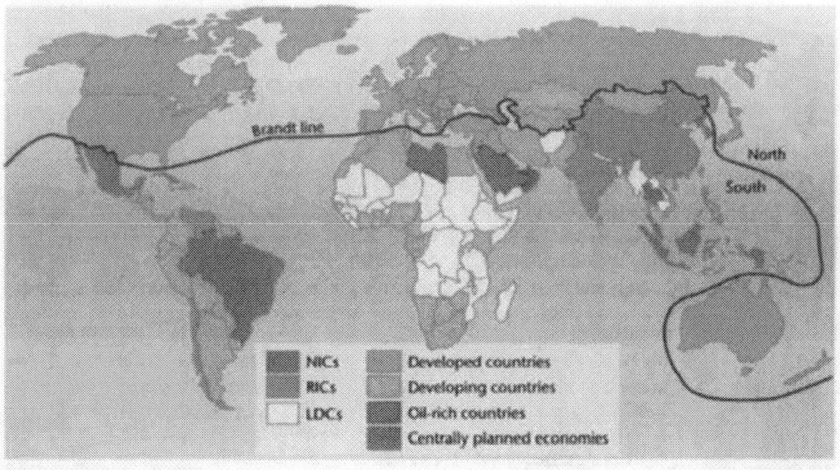

http://developmentandglobalisation.weebly.com

Post-2015 agenda

An alignment to globalized realities

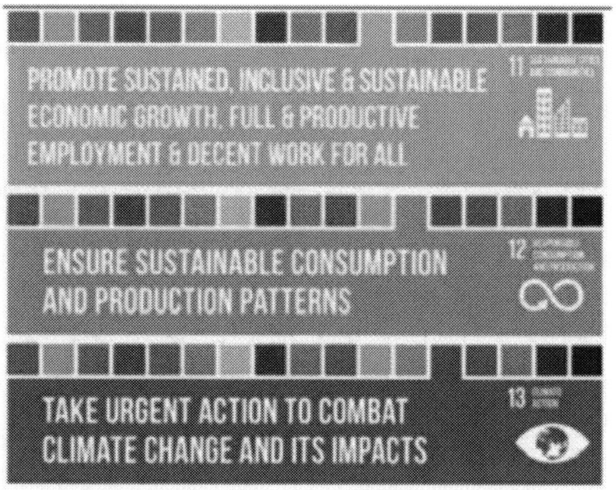

Universality based on common principles

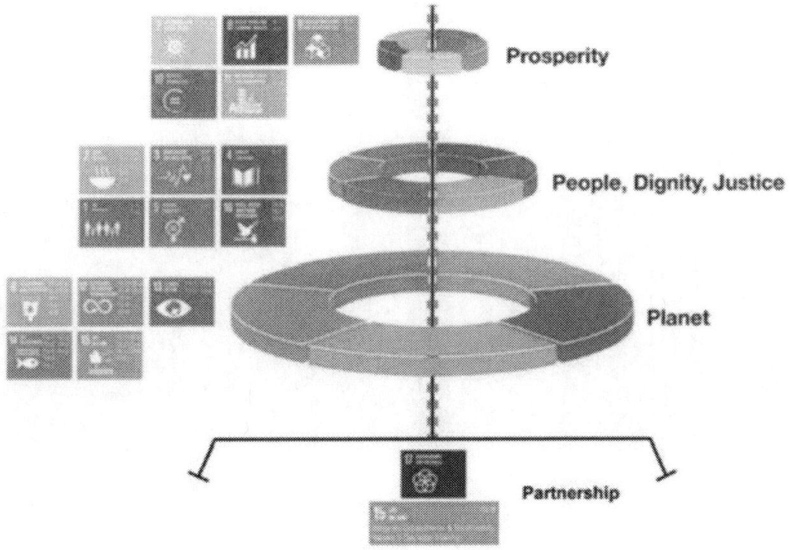

Minal Mistry | Oregon Department of Environmental Quality 17-Dec-19 17

Source: UNDP

Minal Mistry | Oregon Department of Environmental Quality 17-Dec-19 18

SDG and corporate priorities (and omissions)

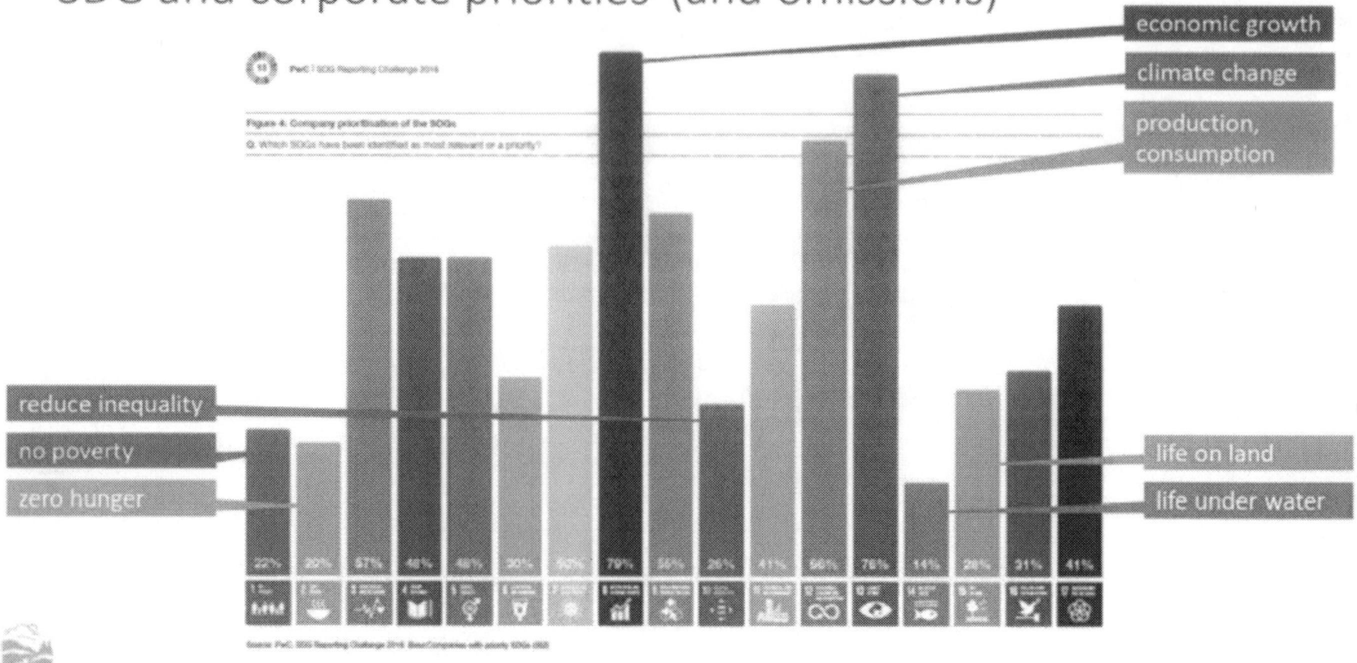

a review of material sustainability frameworks

[https://www.oregon.gov/deq/FilterDocs/mm-matsust.pdf]

sustaining the sustainable

What are we fighting for?

Sustainable Packaging Symposium

16 December 2019

DEQ Minal Mistry | Oregon Department of Environmental Quality

foundation of sustainable development

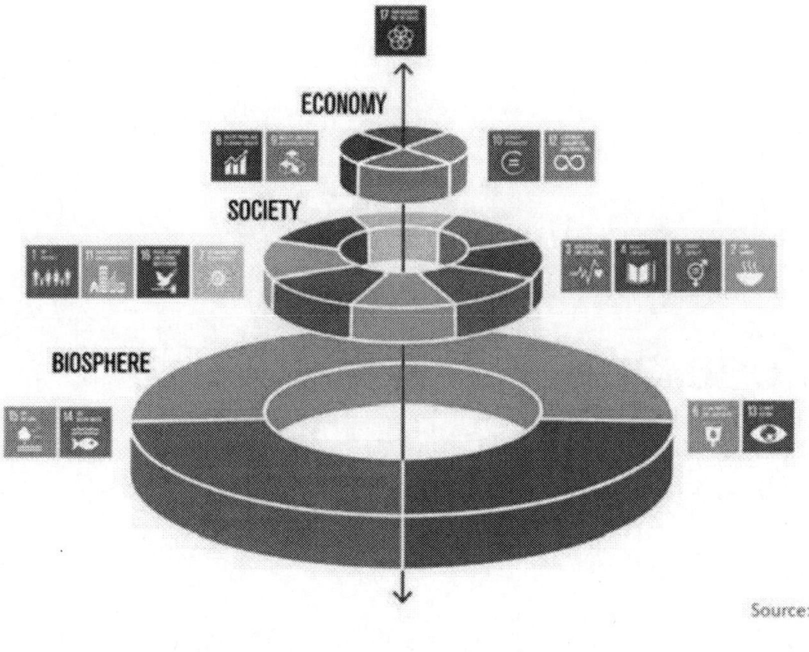

Source: UNDP

DEQ Minal Mistry | Oregon Department of Environmental Quality 17-Dec-19 22

"We are now faced with the fact that tomorrow is today. We are confronted with the fierce urgency of now."

- Martin Luther King Jr.

opprobrium

op·pro·bri·um: harsh criticism or censure

/əˈprōbrēəm/

synopsis

using a wide angle lens

common deep rooted beliefs

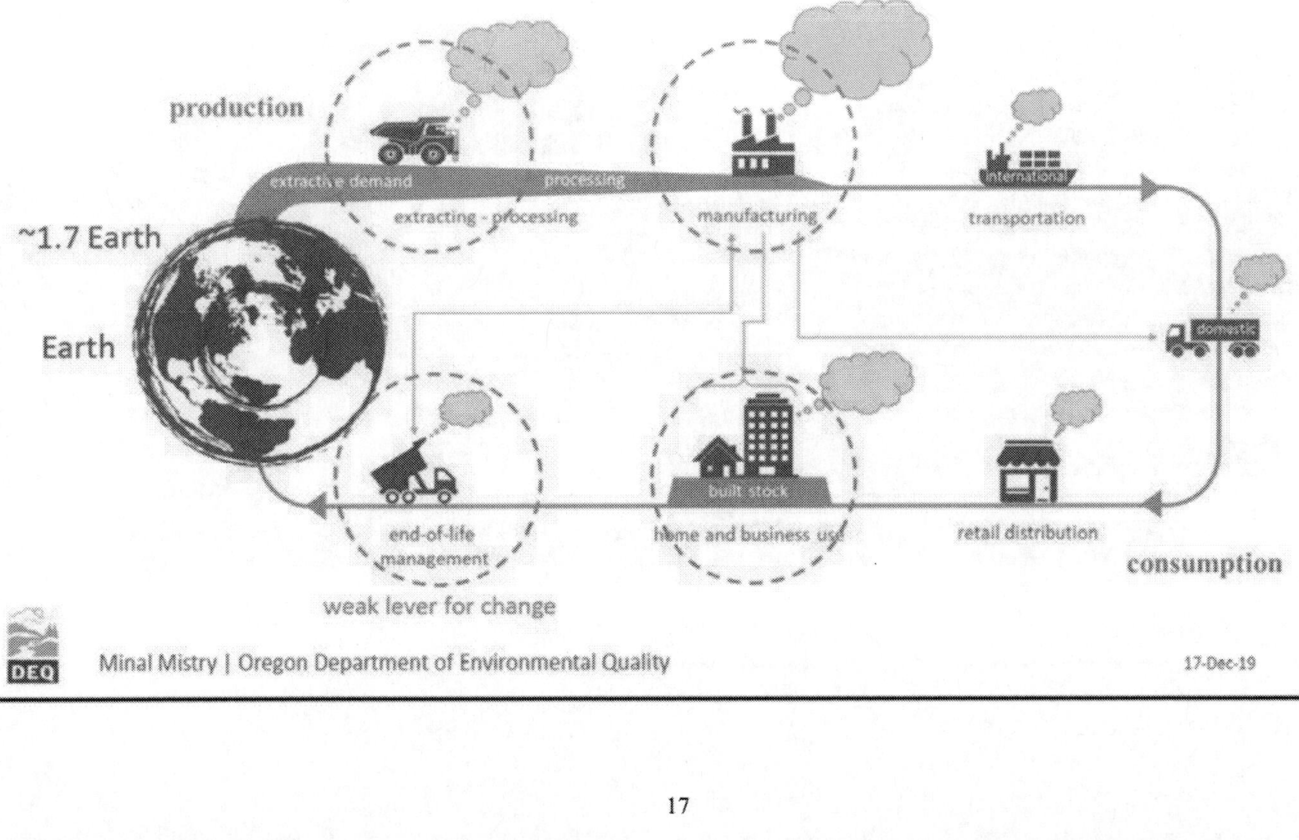

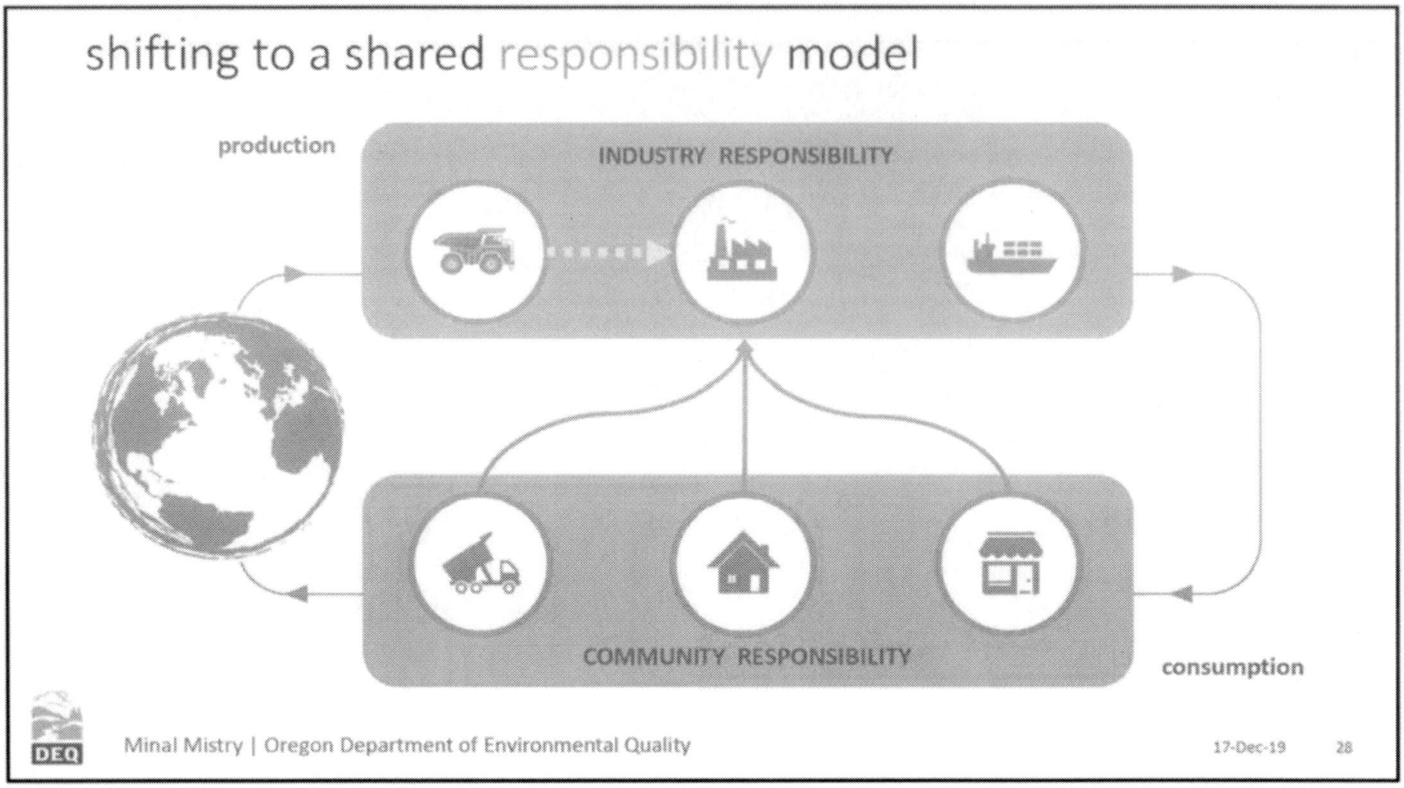
shifting to a shared responsibility model
production
INDUSTRY RESPONSIBILITY
COMMUNITY RESPONSIBILITY
consumption
DEQ
Minal Mistry | Oregon Department of Environmental Quality
17-Dec-19
28

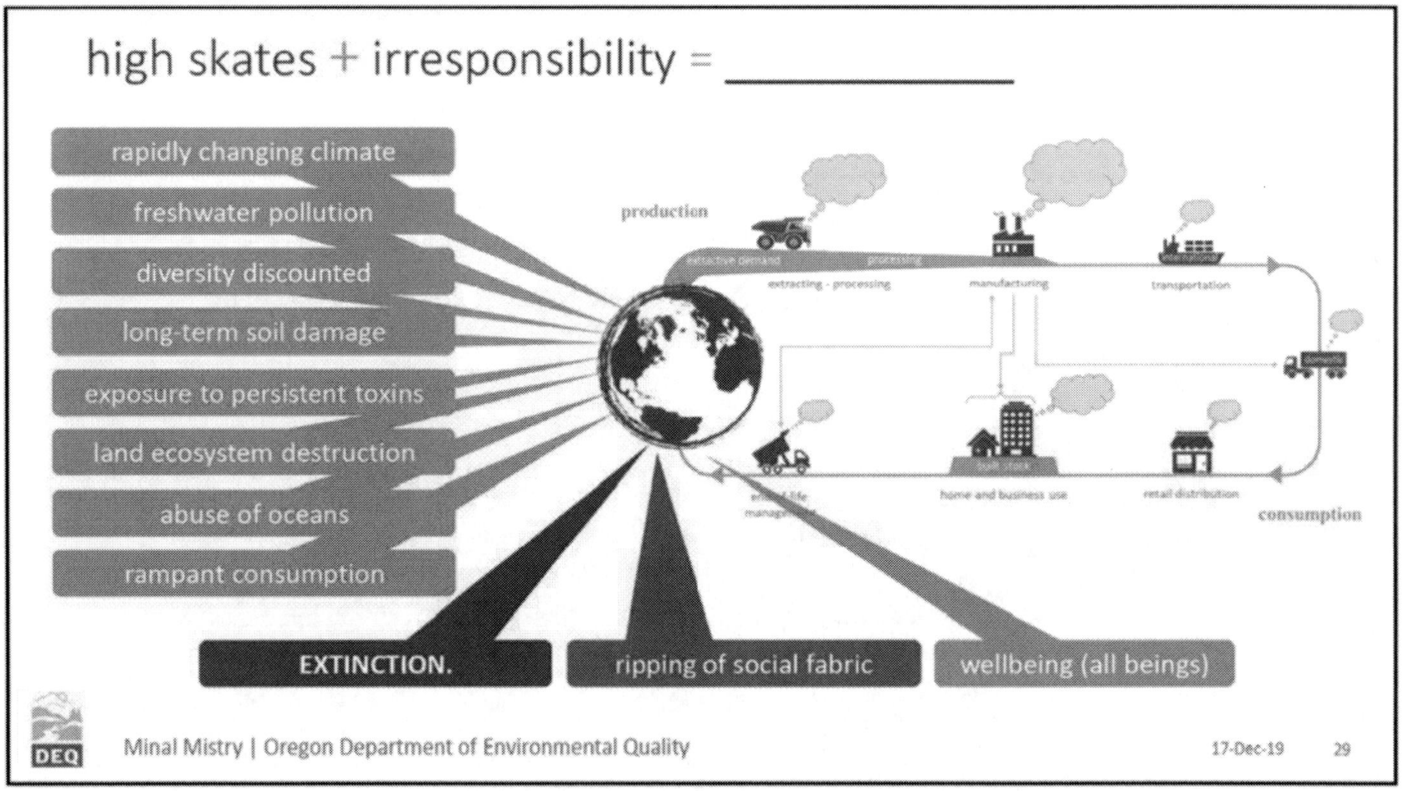
high skates + irresponsibility = ____________
rapidly changing climate
freshwater pollution
diversity discounted
long-term soil damage
exposure to persistent toxins
land ecosystem destruction
abuse of oceans
rampant consumption
production
extracting - processing
manufacturing
transportation
home and business use
retail distribution
consumption
EXTINCTION.
ripping of social fabric
wellbeing (all beings)
DEQ
Minal Mistry | Oregon Department of Environmental Quality
17-Dec-19
29

sustaining wellbeing

sustainable human development

Minal Mistry | Oregon Department of Environmental Quality 17-Dec-19 30

prelude to sustainable development

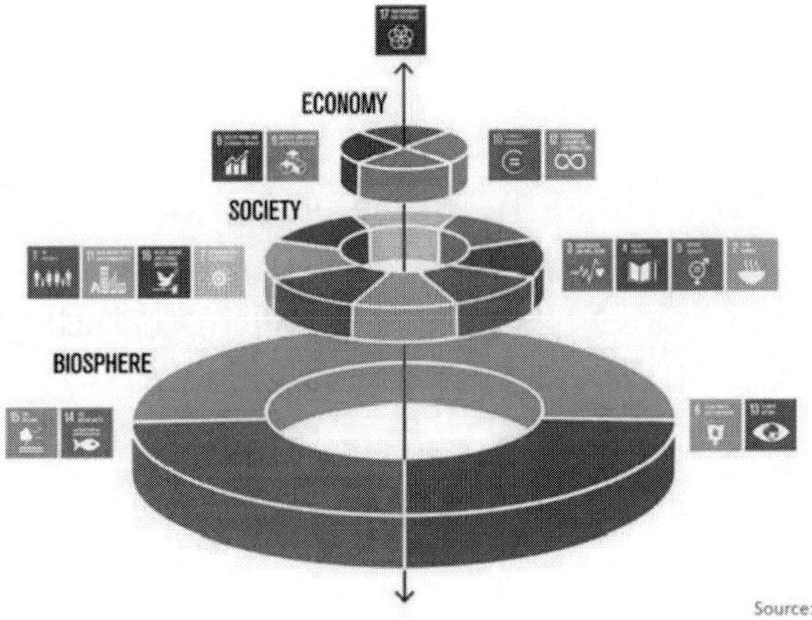

Source: UNDP

Minal Mistry | Oregon Department of Environmental Quality 17-Dec-19 31

'Let us take the decision to confront the questions we have and explore our mutual curiosity, in order to throw light on our similarities without hiding our differences.'

- Jean-Francois Revel

Minal Mistry | Oregon Department of Environmental Quality 17-Dec-19 33

sustainability constructs

frameworks of organization

outlooks of different frameworks

lean thinking
- Efficiency
- Elimination of waste
- Value creation
- Output velocity
- Process simplification

pollution prevention
- Source reduction
- Hazard reduction
- Exposure control

zero waste
- Traditional waste hierarchy
- Elimination of production waste
- Redesign for reuse
- Waste diversion from landfill

circular economy
- Closing material loops
- Upstream focus – redesign, repair, reuse, remanufacturing
- New business models to decouple growth from materials
- Design out externalities

sustainable materials management
- Life cycle thinking
- Impact quantification
- Upstream focus – redesign for repair, reuse, remanufacturing, recycling
- Material stewardship
- Toxics reduction/elimination

sustainable development goals*
- Human wellbeing
- Ecological stewardship
- Economic equity

*not a sustainability framework

who is using these frameworks

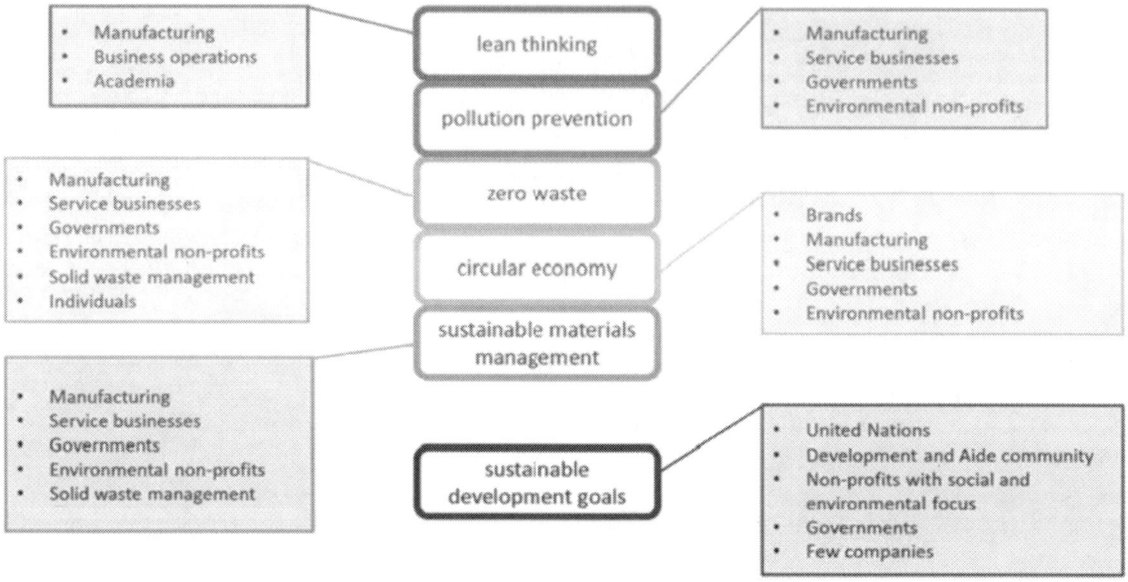

DEQ Minal Mistry | Oregon Department of Environmental Quality 17-Dec-19 36

sustainability focus by frameworks

Framework	Principle Sustainability Focus Area		
	Biosphere (Environment)	Society	Economy
Lean Thinking (lean)			Business operations and production
Pollution Prevention (P2)	Avoidance of pollution via proactive preventive beyond compliance approaches		
Zero Waste (ZW)	Solid waste reduction/avoidance primarily via recycling and composting		Secondary (recycled) materials market
Circular Economy (CE)			Business growth via material circularity (recovery, closed loops, and redesign)
Sustainable Materials Management (SMM)	Reduction of environmental burdens associated with making and using materials		Secondary (recycled) materials market

DEQ Minal Mistry | Oregon Department of Environmental Quality 17-Dec-19 37

lean thinking

optimize production via systematically eliminating waste to meet customer need

links between lean thinking and 'green'

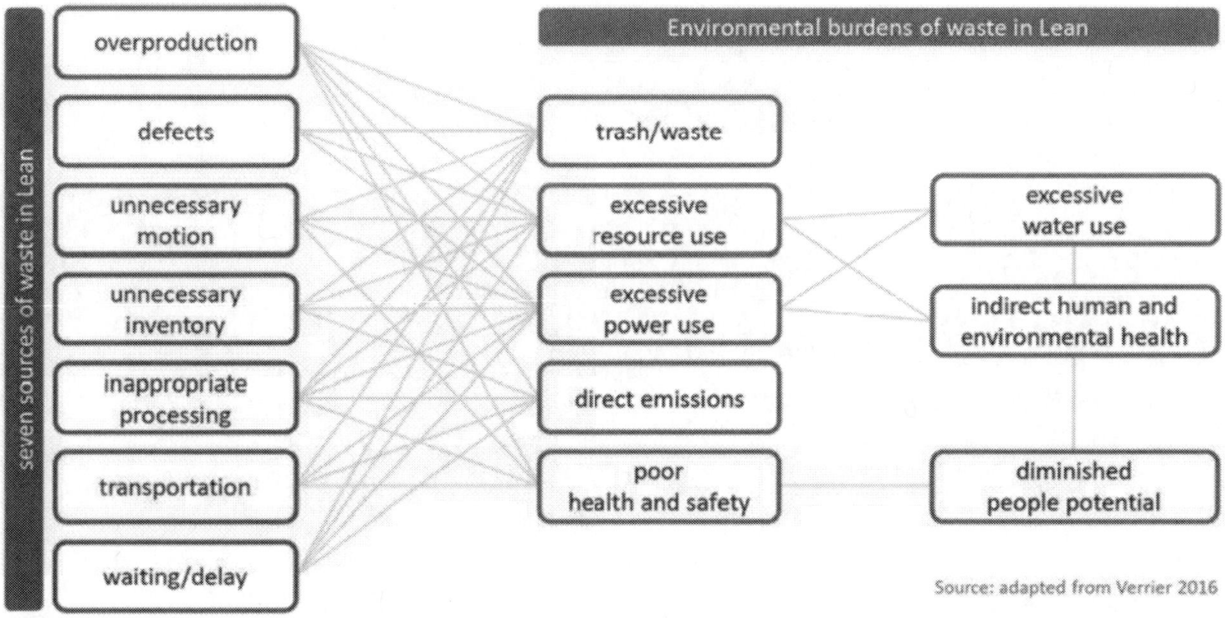

Source: adapted from Verrier 2016

Lean in theory

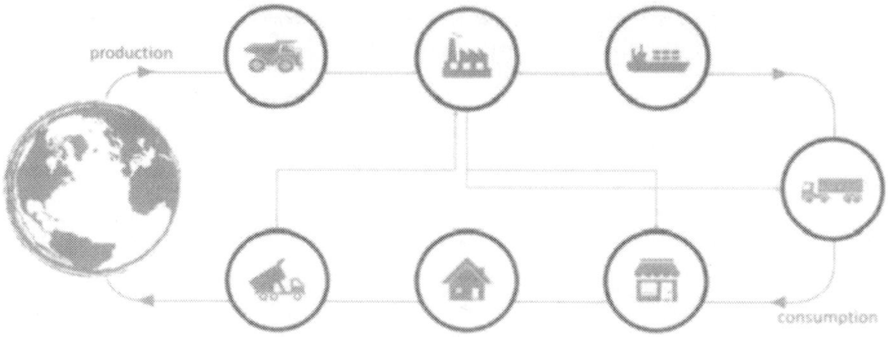

Lean in practice

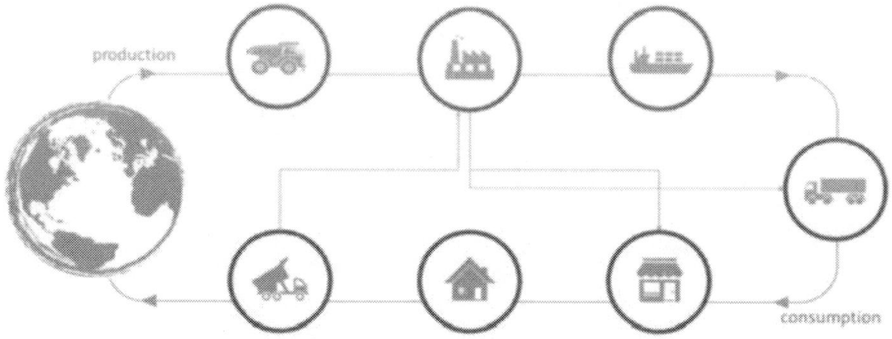

solid circle = fully implemented in stage,
dotted circle = partially implemented in stage

pollution prevention

reduce, eliminate, prevent pollution at the source

pollution prevention (P2)

P2 hierarchy

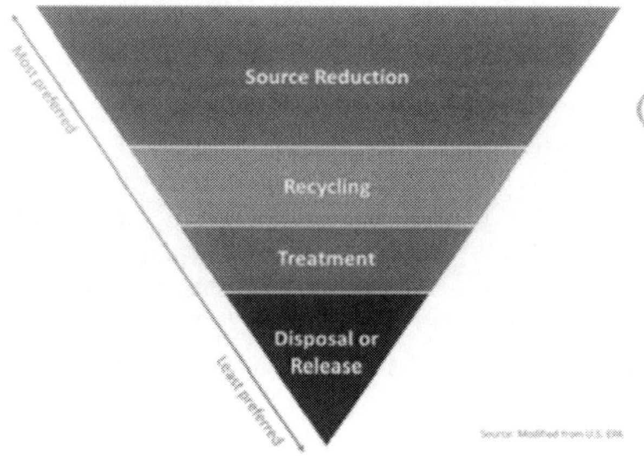

core principles

1. Conserving and protecting natural resources
2. Reducing both financial costs (waste management and cleanup) and environmental costs (health problems and environmental damage)
3. More efficient use of financial resources through more efficient production in industry

P2 in theory

P2 in practice

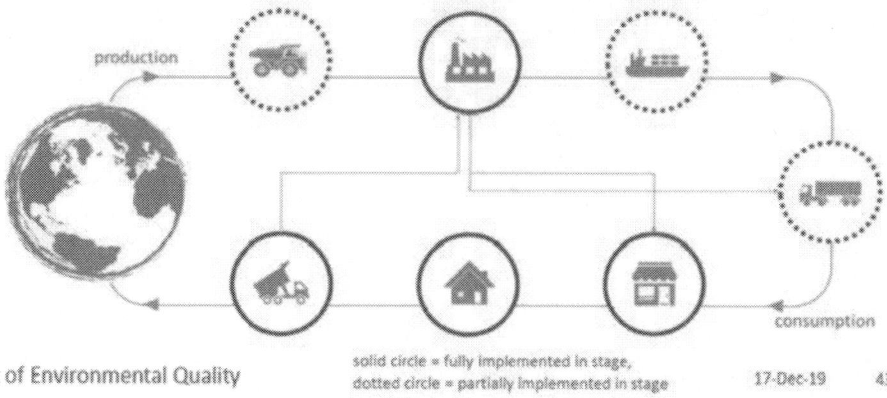

solid circle = fully implemented in stage,
dotted circle = partially implemented in stage

zero waste

aspires to totally eliminate the construct of garbage

Minal Mistry | Oregon Department of Environmental Quality 17-Dec-19 44

zero waste (ZW)

ZW hierarchy

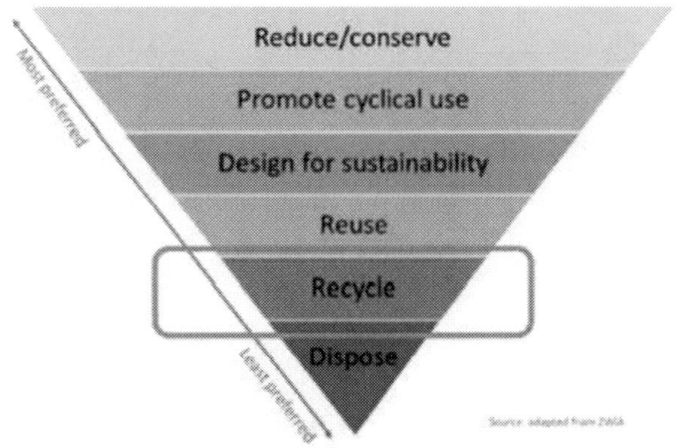

core principles

1. Commitment to the triple bottom line
2. Use Precautionary Principle
3. No waste to landfill or incineration
4. Responsibility: takeback products and packaging
5. Buy reused, recycled and compostable
6. Prevent pollution and reduce waste
7. Highest and best use
8. Use economic incentives for customers, workers and suppliers
9. Non-toxic production, reuse and recycling processes

Minal Mistry | Oregon Department of Environmental Quality 17-Dec-19 45

ZW in theory

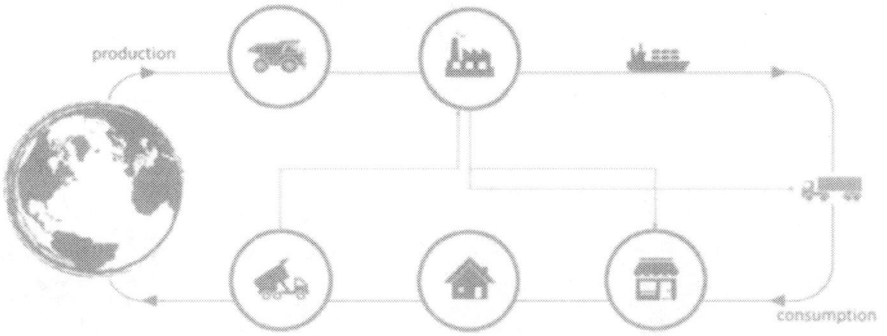

ZW in practice

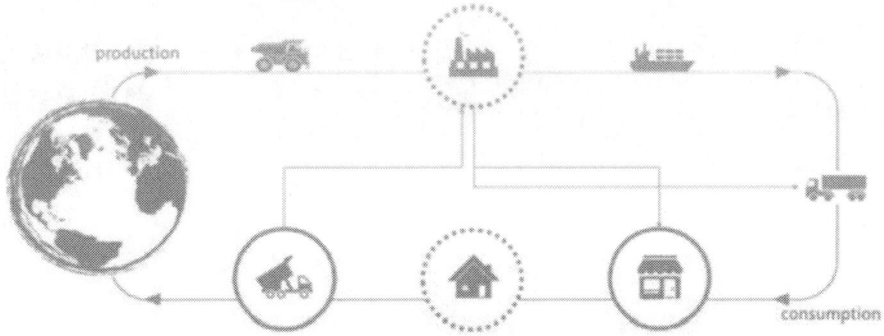

solid circle = fully implemented in stage,
dotted circle = partially implemented in stage

circular economy

regenerative system where resource inputs, waste, emissions, and energy leakage are minimized by slowing, closing, and narrowing material and energy loops

circular economy (CE)

CE hierarchy

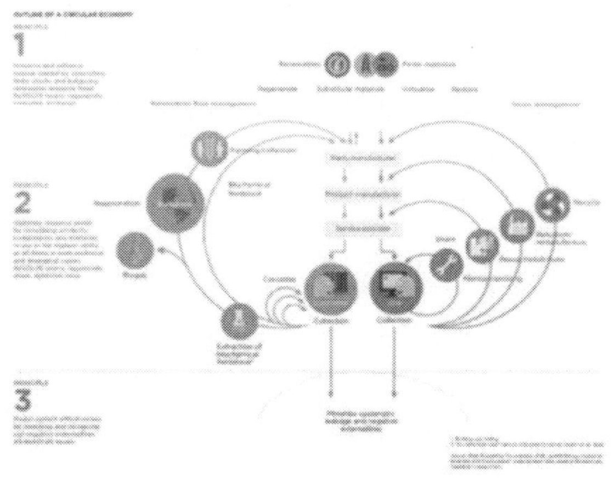

core principles

1. Preserve and enhance natural capital
2. Circulating products, components, and materials at the highest utility
3. Reveal and design out negative externalities

Minal Mistry | Oregon Department of Environmental Quality 17-Dec-19 48

CE in theory

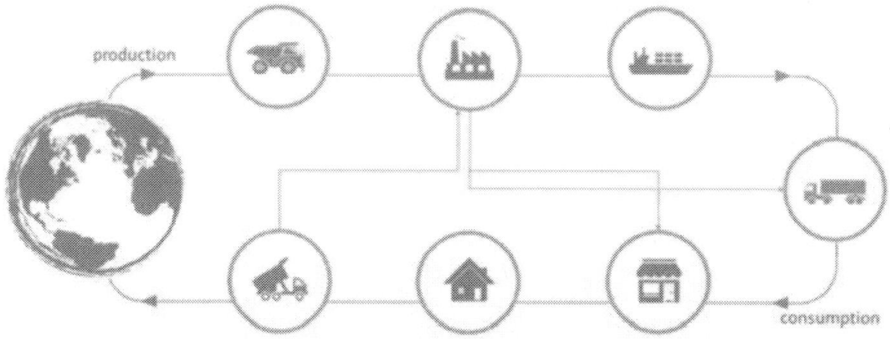

CE in practice

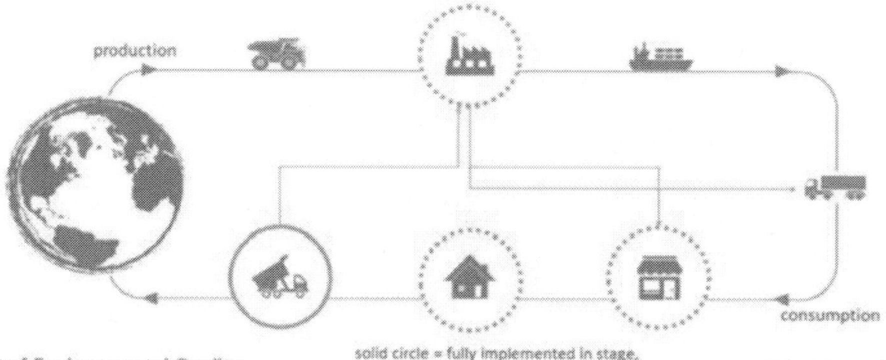

solid circle = fully implemented in stage,
dotted circle = partially implemented in stage

Minal Mistry | Oregon Department of Environmental Quality 17-Dec-19 49

sustainable materials management

apply life cycle thinking to prioritize materials use for best and highest options

Minal Mistry | Oregon Department of Environmental Quality 17-Dec-19 50

sustainable materials management (SMM or MM)

SMM flow

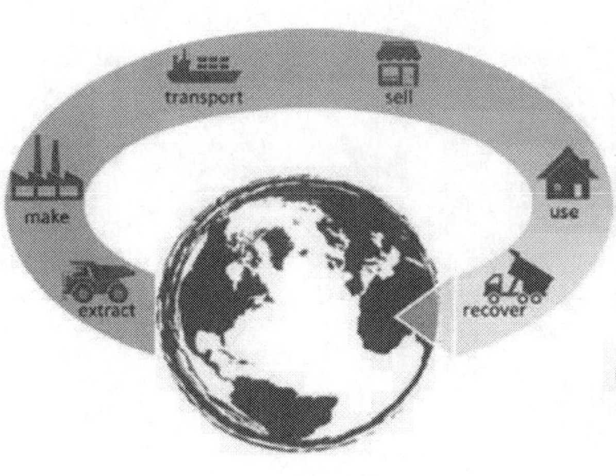

core principles

1. Preserve natural capital
2. Design and manage materials, products and processes using a lifecycle perspective
3. Use the full diversity of policy instruments to stimulate and reinforce sustainable economic, environmental, and social outcomes
4. Engage all parts of society to take active, ethically-based responsibility for achieving sustainable outcomes

Minal Mistry | Oregon Department of Environmental Quality 17-Dec-19 51

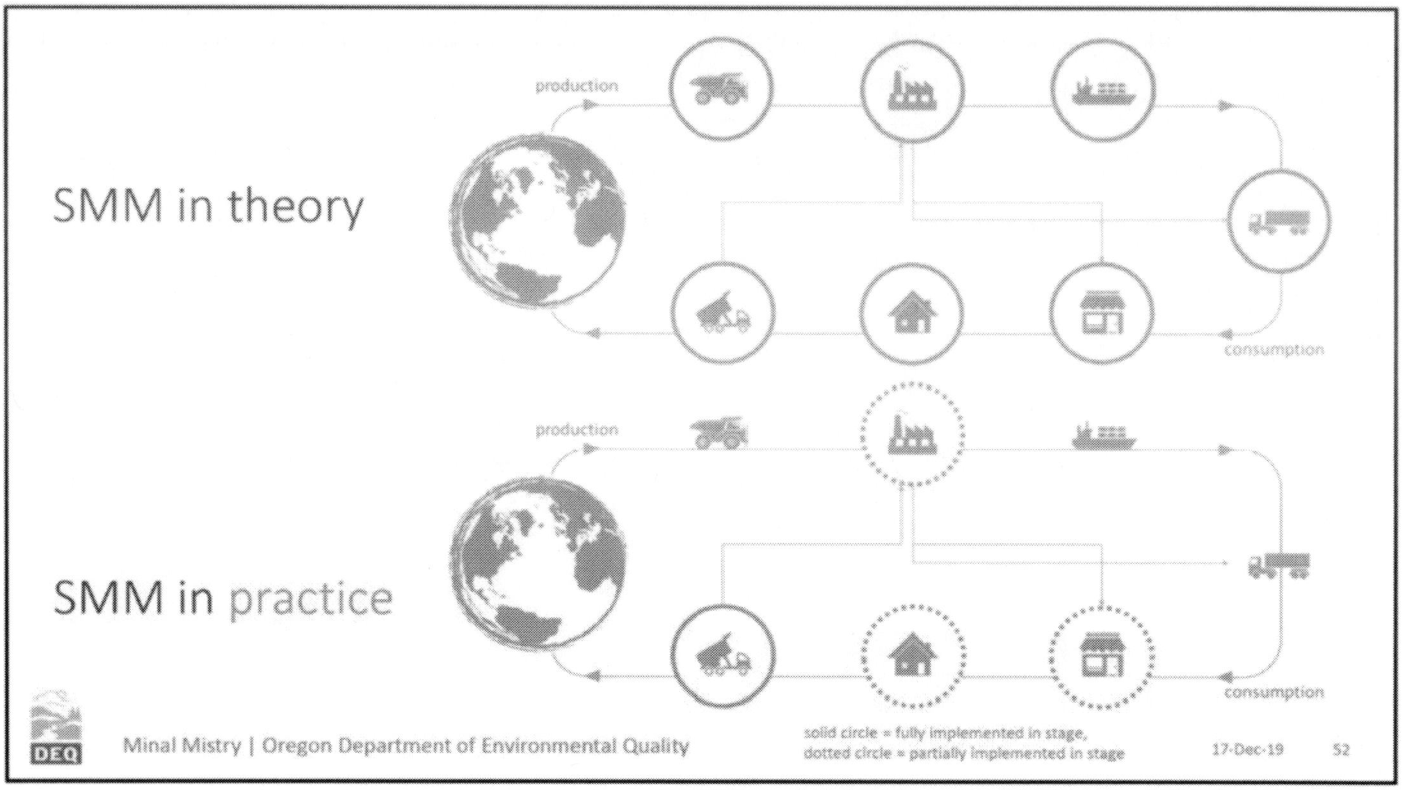
SMM in theory
production
consumption
SMM in practice
production
consumption
Minal Mistry | Oregon Department of Environmental Quality
solid circle = fully implemented in stage,
dotted circle = partially implemented in stage
17-Dec-19
52

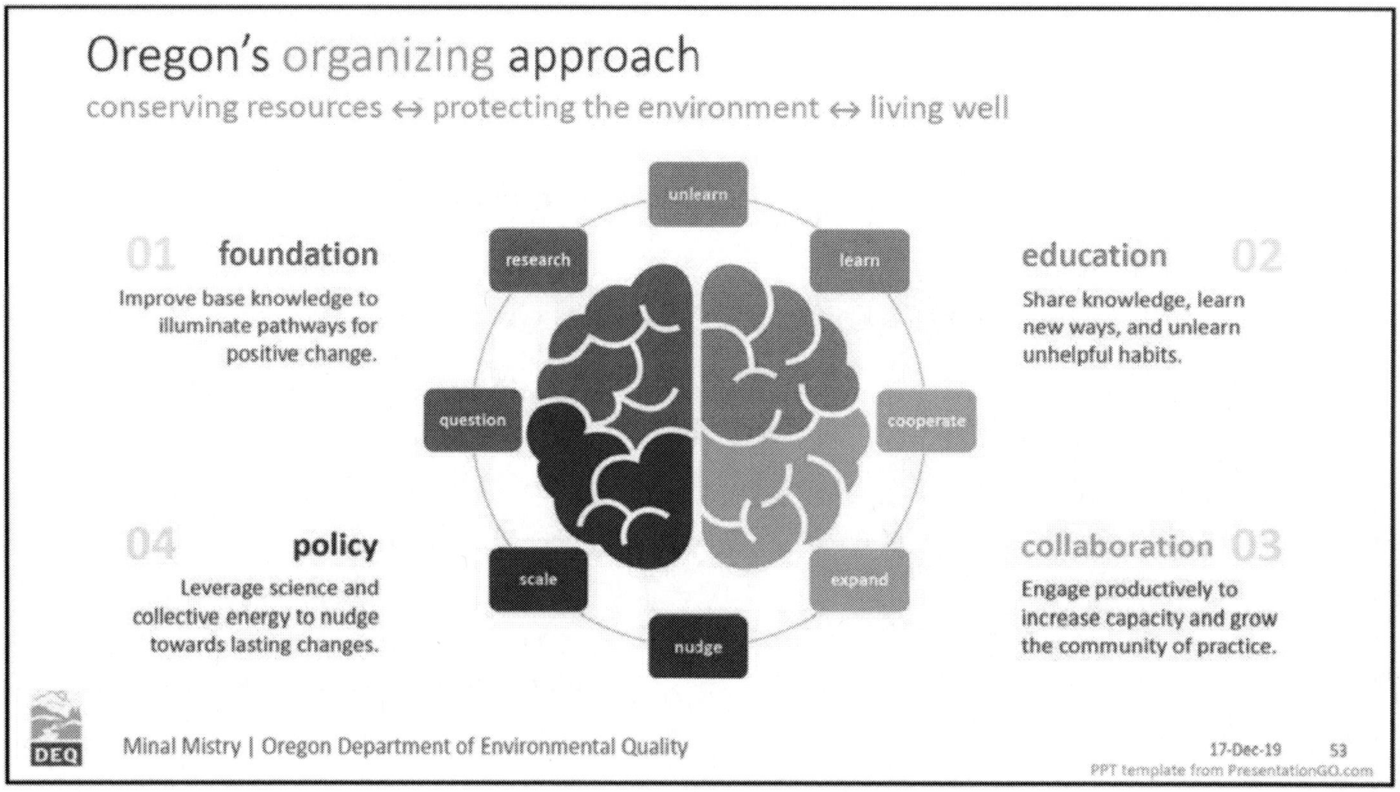
Oregon's organizing approach
conserving resources ↔ protecting the environment ↔ living well
01 foundation
Improve base knowledge to illuminate pathways for positive change.
education 02
Share knowledge, learn new ways, and unlearn unhelpful habits.
unlearn
research
learn
question
cooperate
scale
expand
nudge
04 policy
Leverage science and collective energy to nudge towards lasting changes.
collaboration 03
Engage productively to increase capacity and grow the community of practice.
Minal Mistry | Oregon Department of Environmental Quality
17-Dec-19
53
PPT template from PresentationGO.com

synthesis

action ➲⊂ reaction

Minal Mistry | Oregon Department of Environmental Quality 17-Dec-19 54

the potential of sustainability frameworks

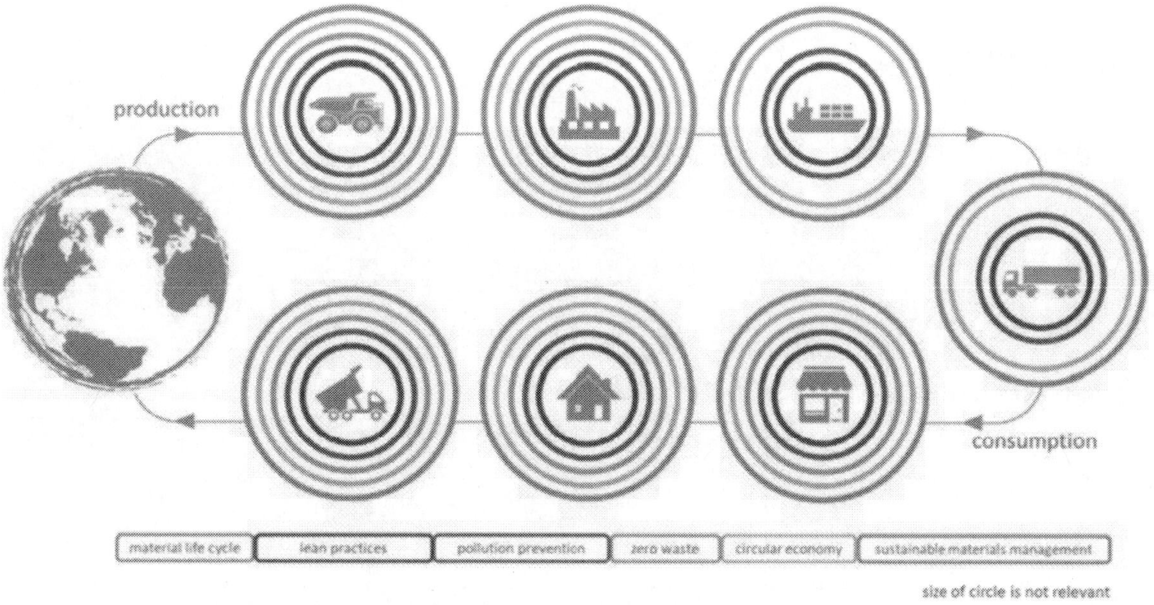

Minal Mistry | Oregon Department of Environmental Quality 17-Dec-19 55

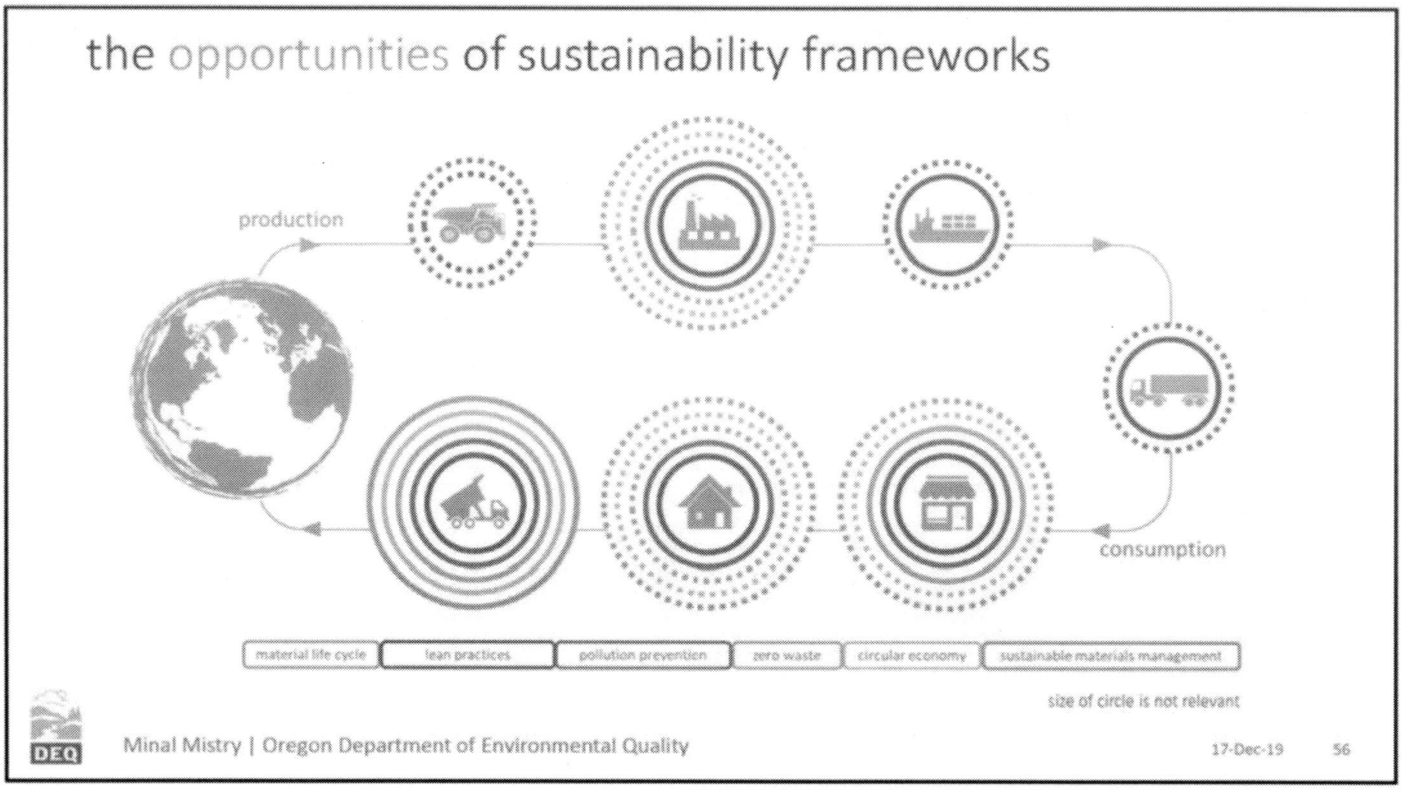
the opportunities of sustainability frameworks
production
consumption
material life cycle
lean practices
pollution prevention
zero waste
circular economy
sustainable materials management
size of circle is not relevant
DEQ
Minal Mistry | Oregon Department of Environmental Quality
17-Dec-19
56

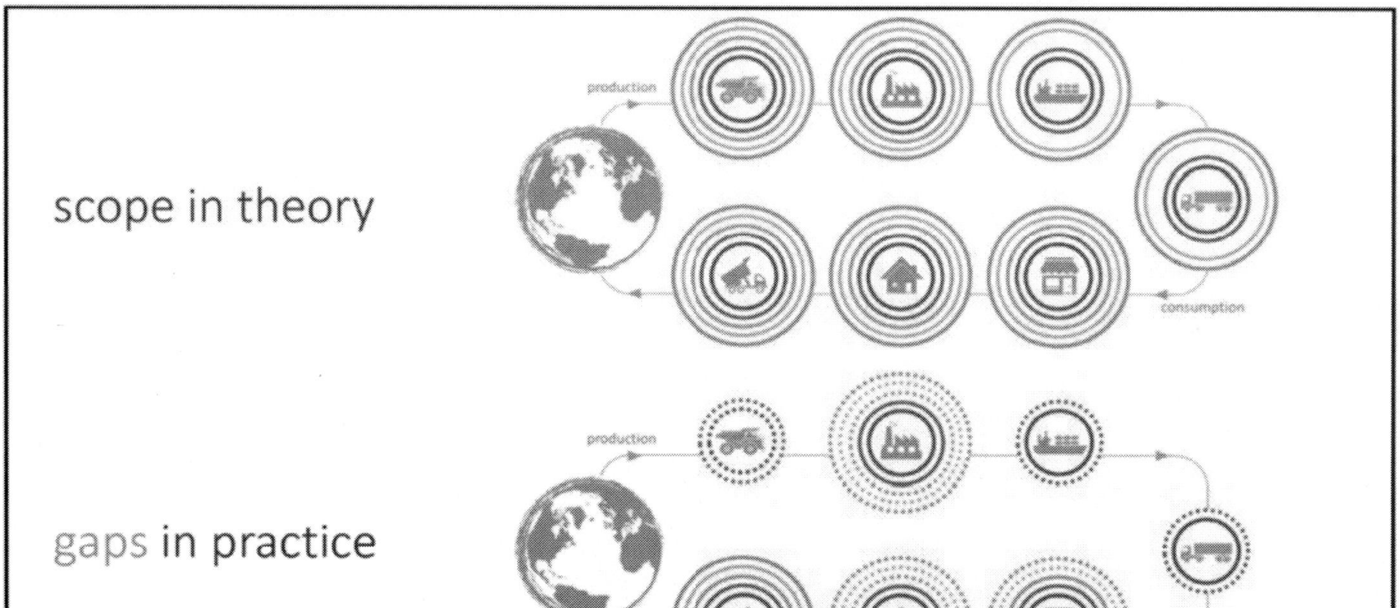
scope in theory
production
consumption
gaps in practice
production
consumption
DEQ
Minal Mistry |
material life cycle
lean practices
pollution prevention
zero waste
circular economy
sustainable materials management
57

connecting the dots

cooperation and collaboration

the overlap between frameworks

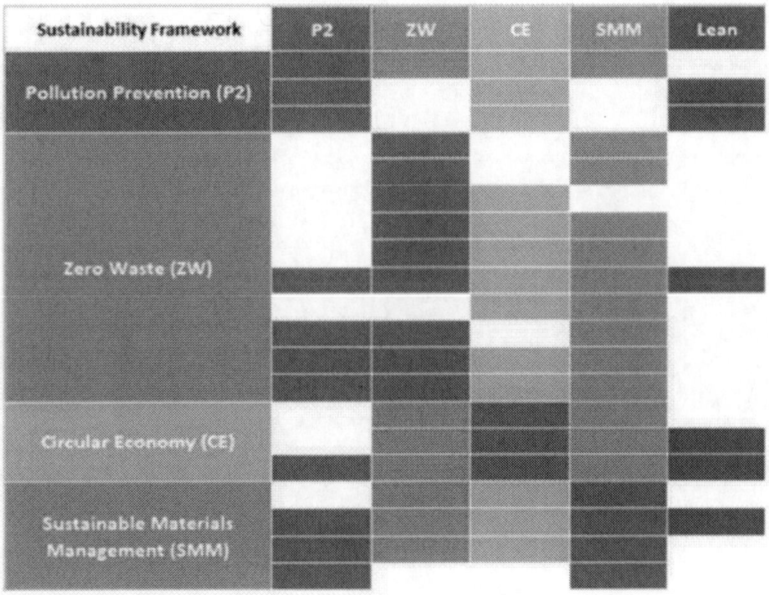

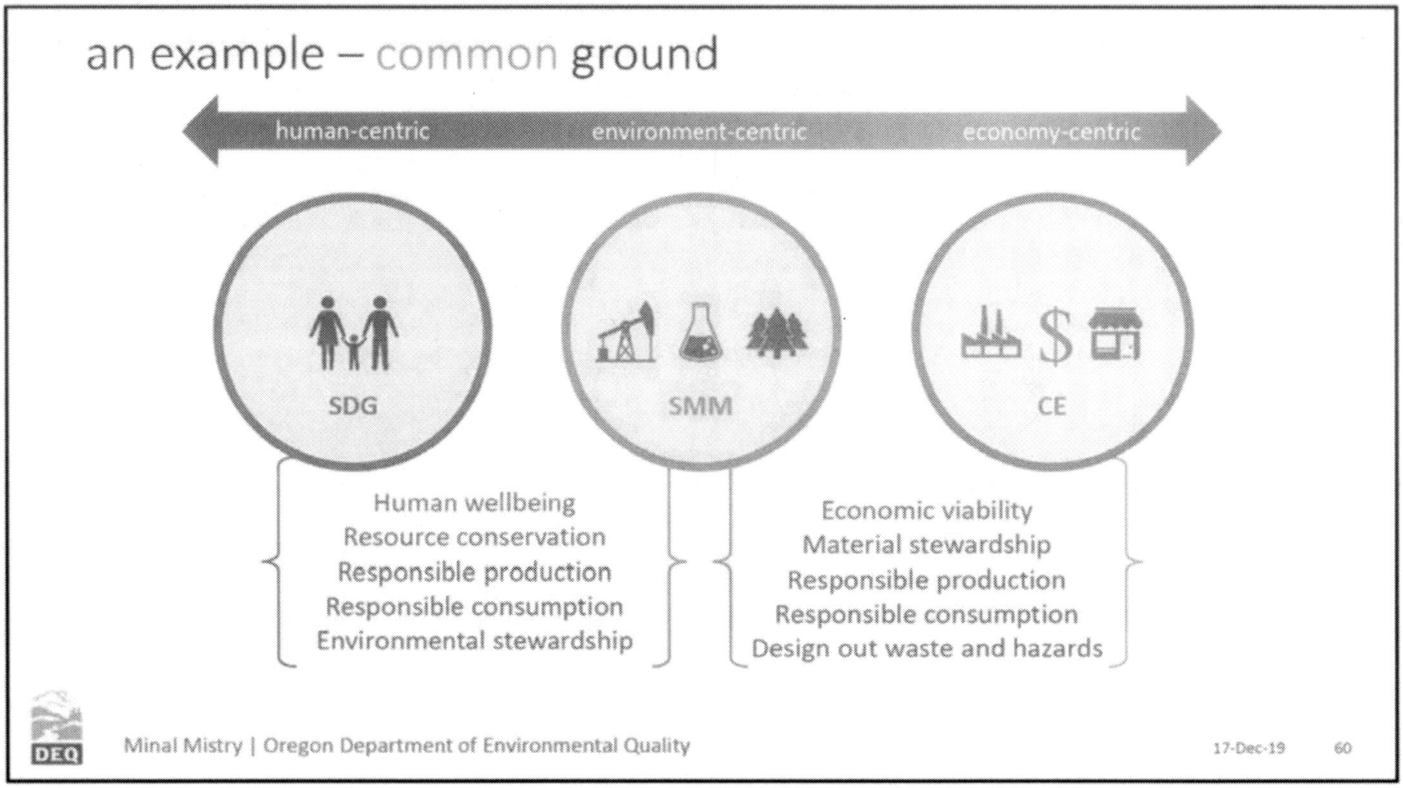
an example – common ground
human-centric
environment-centric
economy-centric
SDG
SMM
CE
Human wellbeing
Resource conservation
Responsible production
Responsible consumption
Environmental stewardship
Economic viability
Material stewardship
Responsible production
Responsible consumption
Design out waste and hazards
DEQ
Minal Mistry | Oregon Department of Environmental Quality
17-Dec-19
60

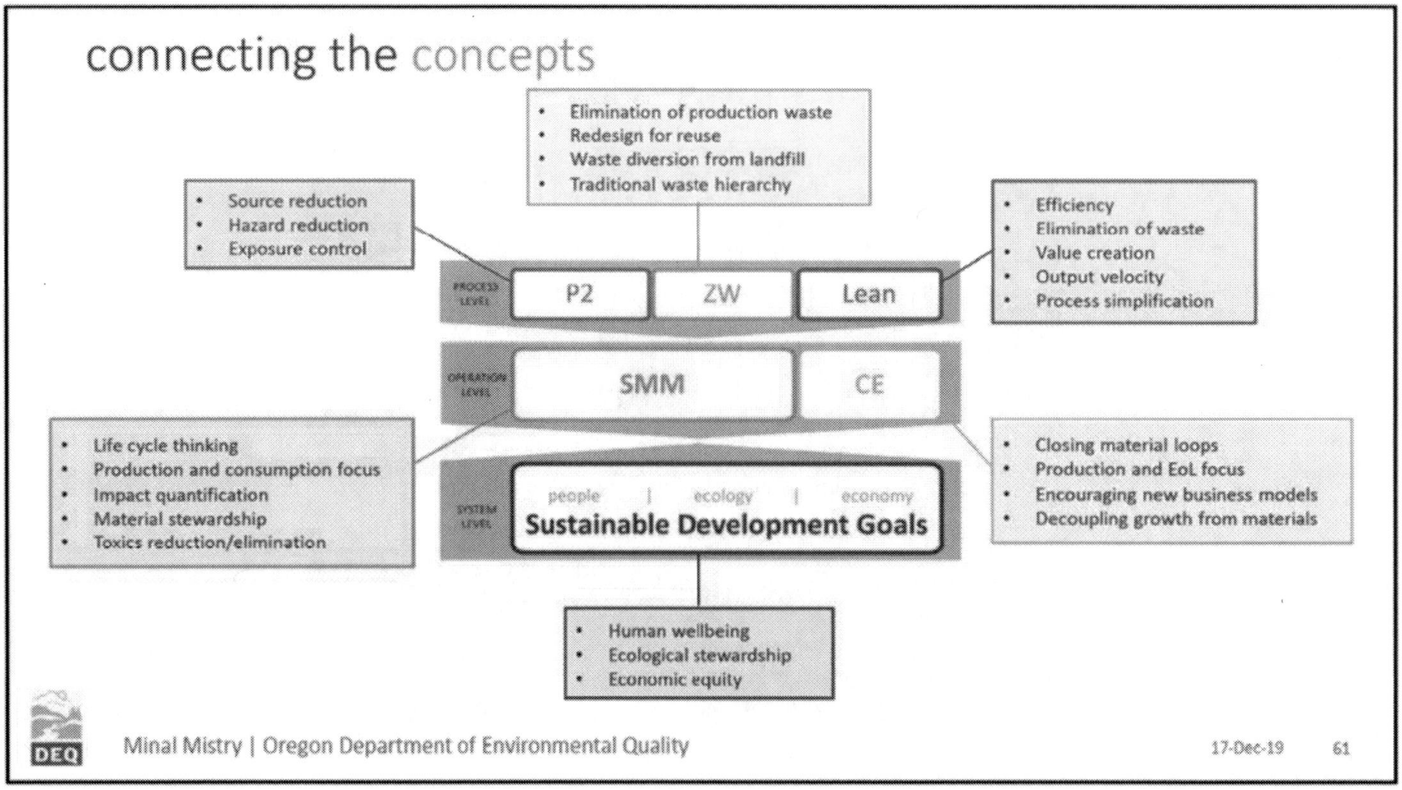
connecting the concepts
Elimination of production waste
Redesign for reuse
Waste diversion from landfill
Traditional waste hierarchy
Source reduction
Hazard reduction
Exposure control
Efficiency
Elimination of waste
Value creation
Output velocity
Process simplification
PROCESS LEVEL
P2
ZW
Lean
OPERATION LEVEL
SMM
CE
Life cycle thinking
Production and consumption focus
Impact quantification
Material stewardship
Toxics reduction/elimination
Closing material loops
Production and EoL focus
Encouraging new business models
Decoupling growth from materials
SYSTEM LEVEL
people | ecology | economy
Sustainable Development Goals
Human wellbeing
Ecological stewardship
Economic equity
DEQ
Minal Mistry | Oregon Department of Environmental Quality
17-Dec-19
61

industrial ecology: existing supporting tools and knowledge

- Material characterizations
- Life cycle assessment (LCA)
- Hazard assessment
- Alternate assessment
- Green chemistry principles
- Industrial symbiosis
- Process evaluation
- Living design
- Biomimicry…

the big development tent

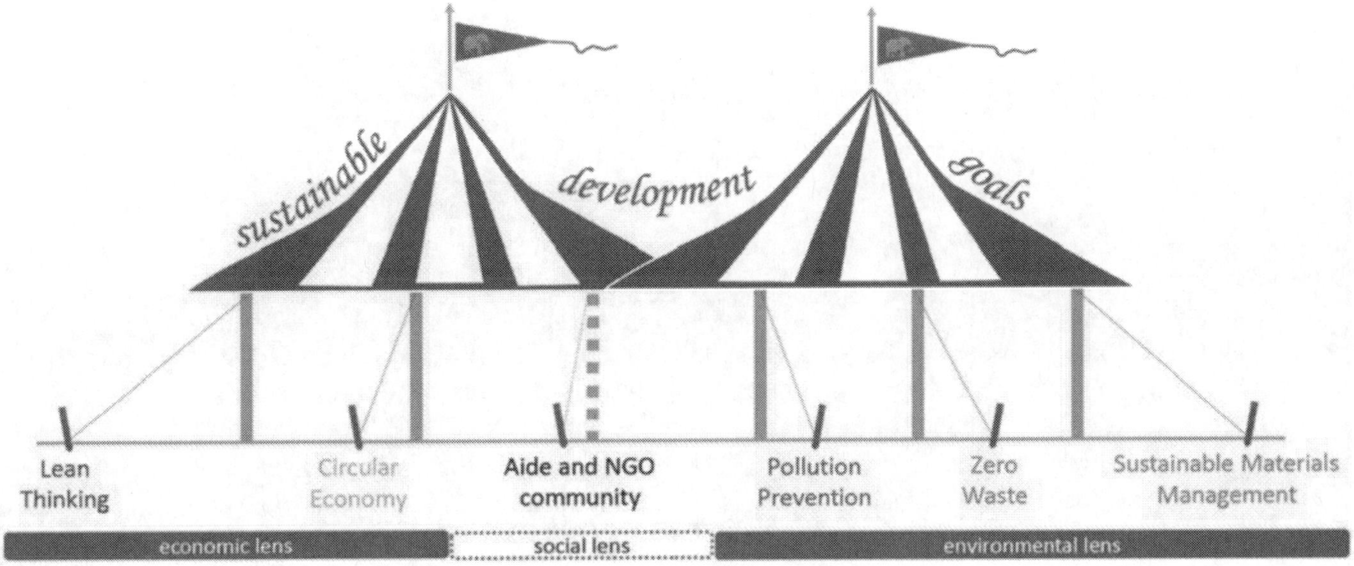

'The pine tree, the leopard, the Platte River, and ourselves – we are at risk together, or we are on our way to a sustainable world together. We are each other's destiny.'

- Mary Oliver from *Upstream*

Minal Mistry | Oregon Department of Environmental Quality 17-Dec-19 64

'In every age there is a turning point, a new way of seeing and asserting the coherence of the world.'

- J. Boronoski from *The Ascent of Man*

materials
management

conserving resources

protecting the environment

living well

minal mistry
minal.mistry@state.or.us
@amistryman

scale matters

opportunities for impact reduction

Sustainable Packaging Symposium

17 December 2019

DEQ Minal Mistry | Oregon Department of Environmental Quality

a systems perspective

life cycle of material flows

Minal Mistry | Oregon Department of Environmental Quality 7-Jan-20 3

material life cycle

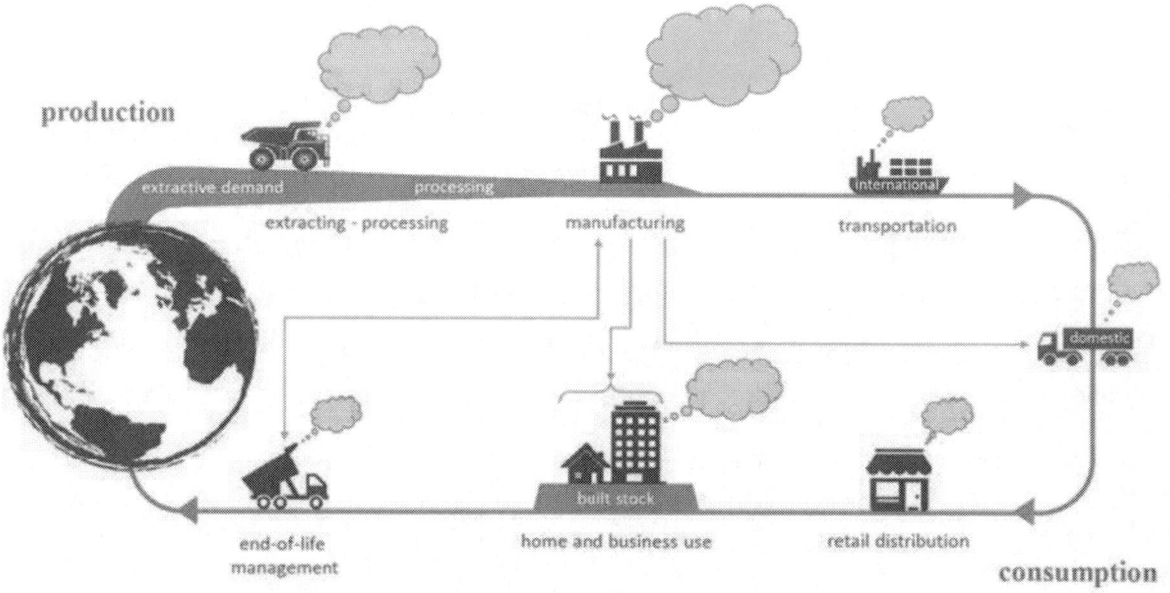

Minal Mistry | Oregon Department of Environmental Quality 7-Jan-20 4

buttress environmental damage

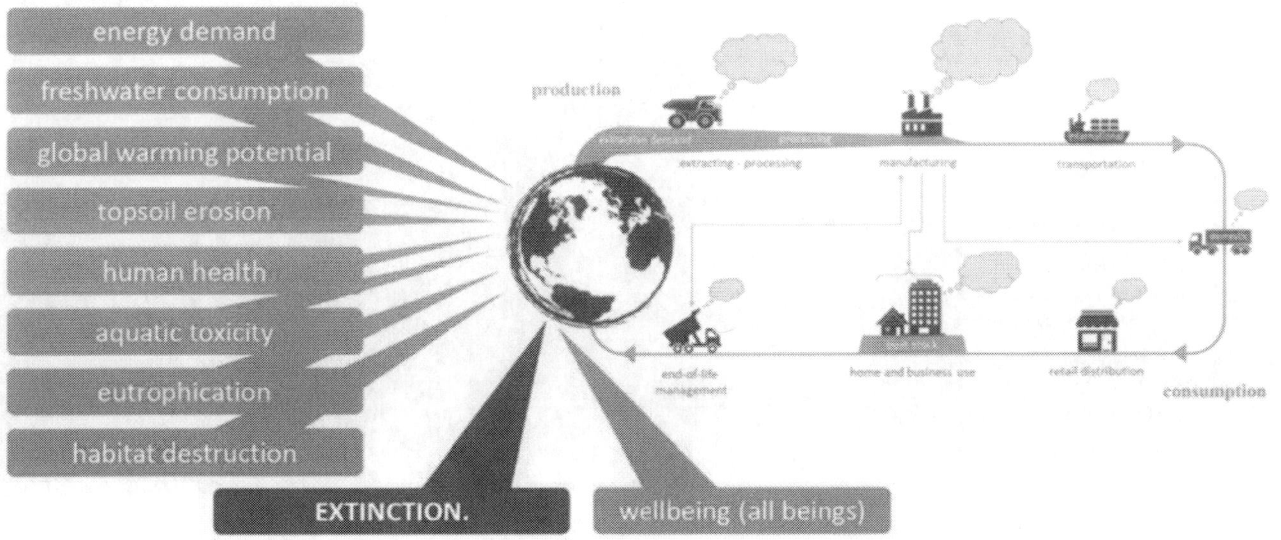

a demand issues

consuming a living organism

GHG inventories in Oregon, 2015

Oregon Economy
(63 MMT CO2e)

Oregon Consumption
(89 MMT CO2e)

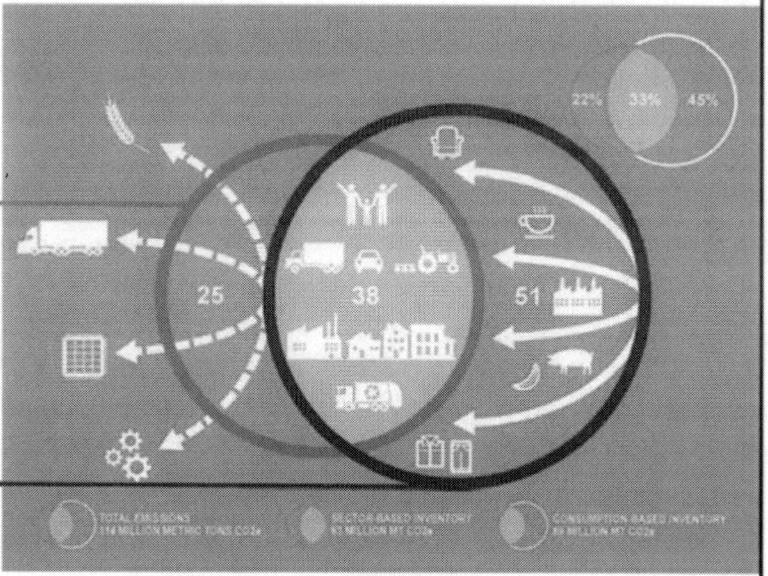

Source: DEQ 2018

Oregon's GHG trajectories, 1990-2015

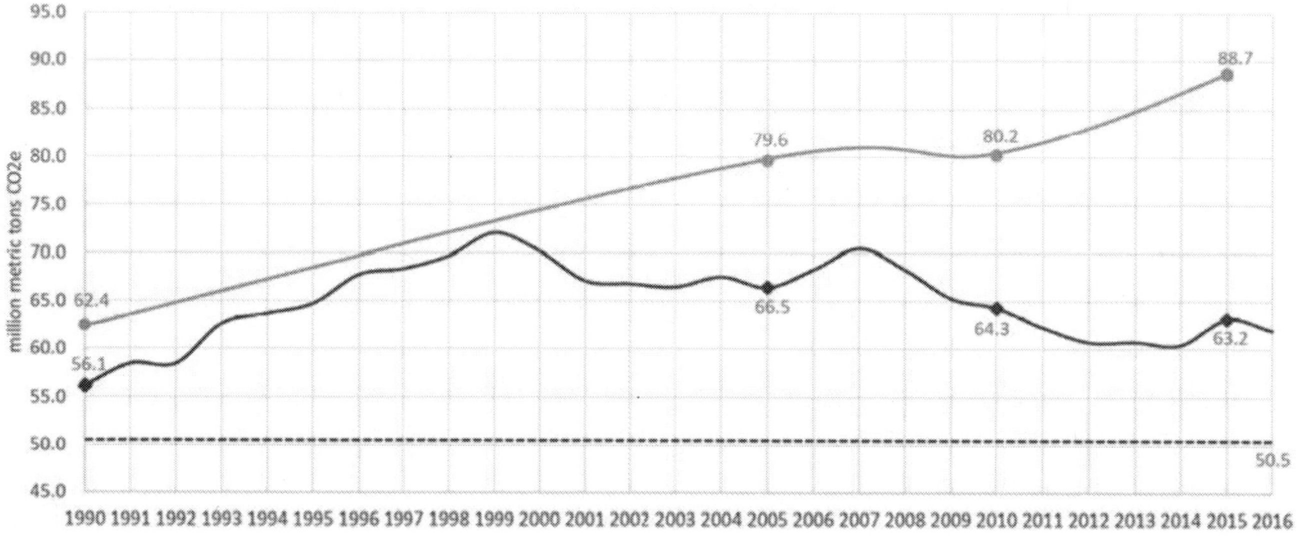

Source: DEQ 2018

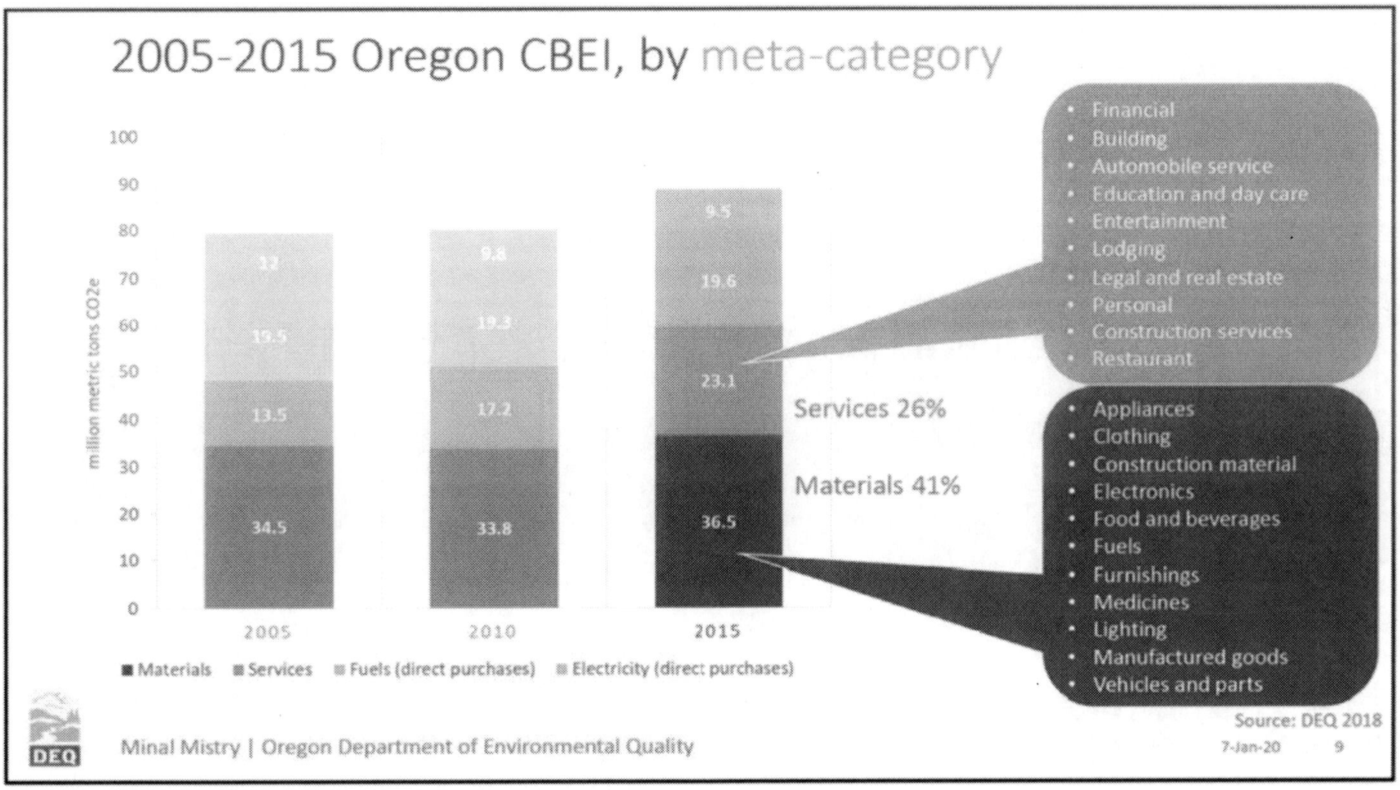
2005-2015 Oregon CBEI, by meta-category
million metric tons CO2e
100
90
80
70
60
50
40
30
20
10
0
2005
2010
2015
12
15.5
13.5
34.5
9.8
19.3
17.2
33.8
9.5
19.6
23.1
36.5
Services 26%
Materials 41%
Materials
Services
Fuels (direct purchases)
Electricity (direct purchases)
Financial
Building
Automobile service
Education and day care
Entertainment
Lodging
Legal and real estate
Personal
Construction services
Restaurant
Appliances
Clothing
Construction material
Electronics
Food and beverages
Fuels
Furnishings
Medicines
Lighting
Manufactured goods
Vehicles and parts
Source: DEQ 2018
DEQ
Minal Mistry | Oregon Department of Environmental Quality
7-Jan-20
9

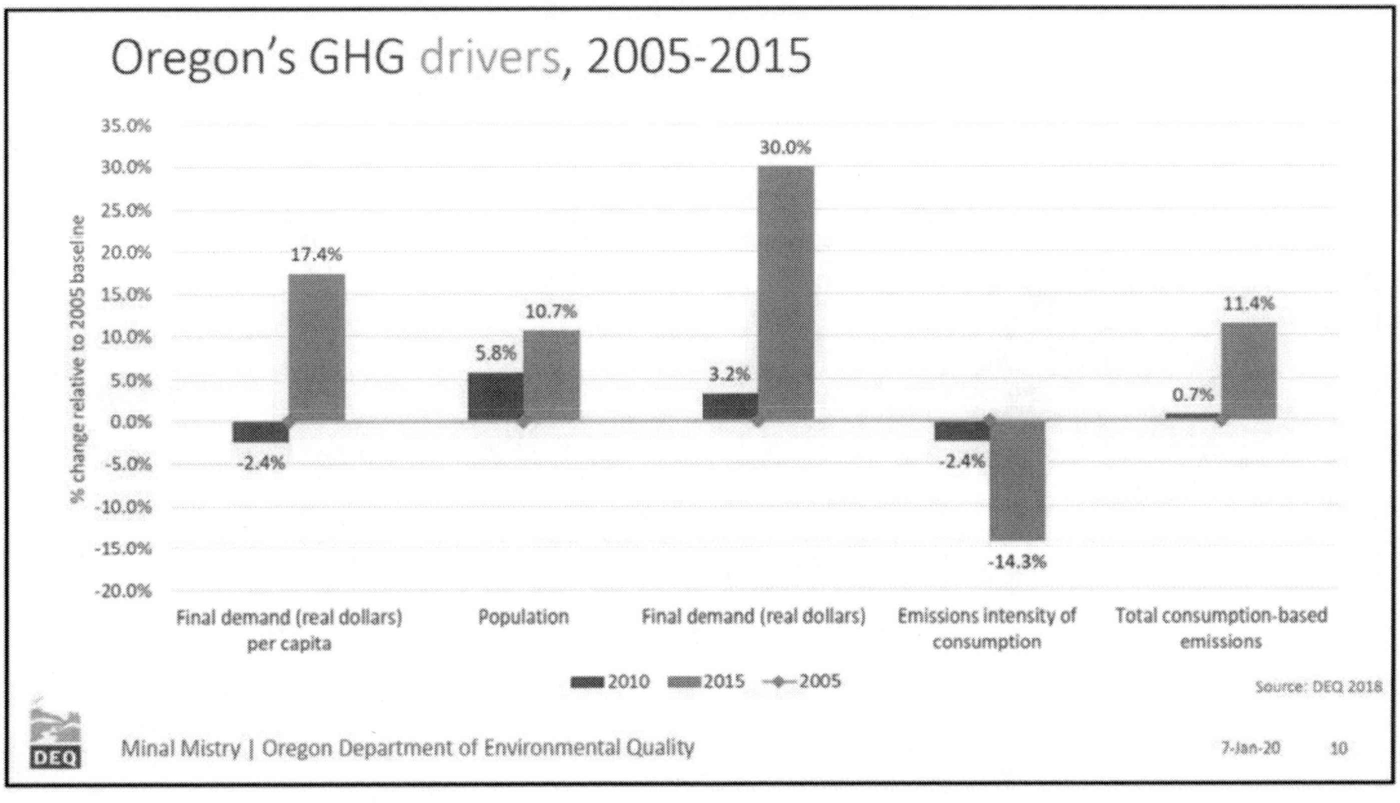
Oregon's GHG drivers, 2005-2015
% change relative to 2005 baseline
35.0%
30.0%
25.0%
20.0%
15.0%
10.0%
5.0%
0.0%
-5.0%
-10.0%
-15.0%
-20.0%
-2.4%
17.4%
5.8%
10.7%
3.2%
30.0%
-2.4%
-14.3%
0.7%
11.4%
Final demand (real dollars) per capita
Population
Final demand (real dollars)
Emissions intensity of consumption
Total consumption-based emissions
2010
2015
2005
Source: DEQ 2018
DEQ
Minal Mistry | Oregon Department of Environmental Quality
7-Jan-20
10

Takeaway #1

Consumption is a root driver of environmental damage.

deeply held material myths

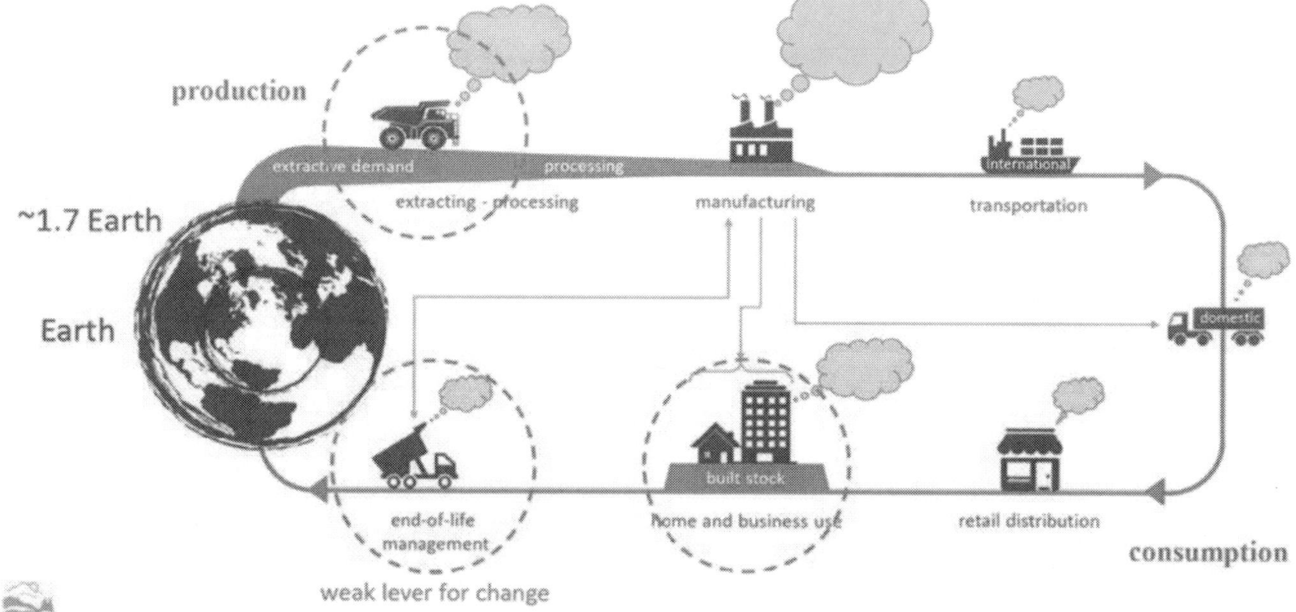

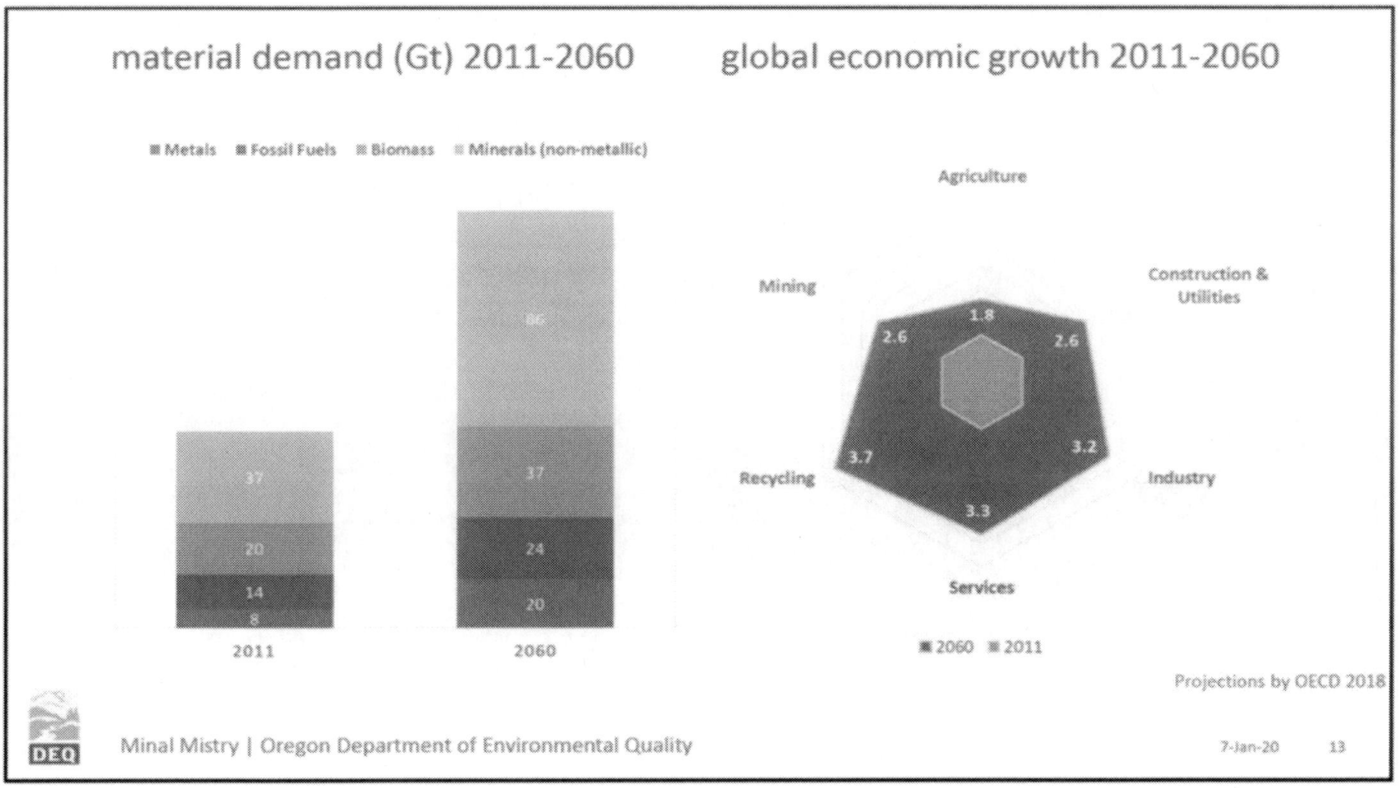
material demand (Gt) 2011-2060
global economic growth 2011-2060
Metals
Fossil Fuels
Biomass
Minerals (non-metallic)
86
37
37
20
24
14
20
8
2011
2060
Agriculture
Construction & Utilities
Industry
Services
Recycling
Mining
1.8
2.6
3.2
3.3
3.7
2.6
2060
2011
Projections by OECD 2018
DEQ
Minal Mistry | Oregon Department of Environmental Quality
7-Jan-20
13

extractive demand
one
...70 years...
one.7
...40 years...
three.5
pre-1950
thriving
present
struggling
~2060
dying
Source: Mistry 2019
DEQ
Minal Mistry | Oregon Department of Environmental Quality
7-Jan-20
14

shifting to a shared responsibility mindset

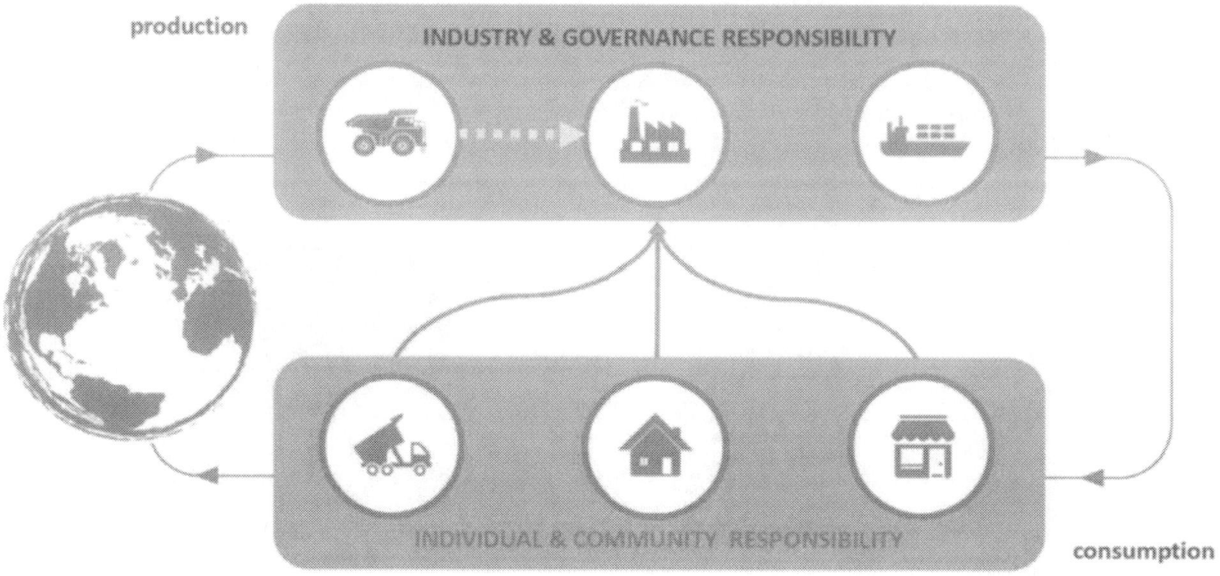

Source: Mistry 2019

packaging attributes

complexity oversimplified

research question

How well do popular material attributes guide us towards reduced environmental impact options?

life cycle impacts and material attributes

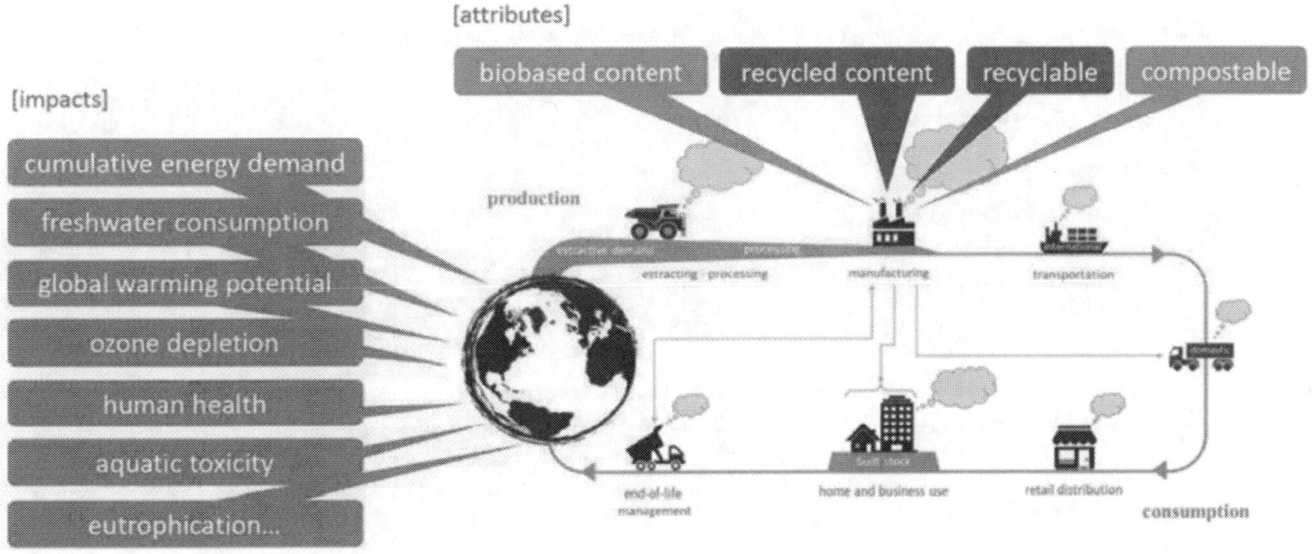

evaluation: sample calculation

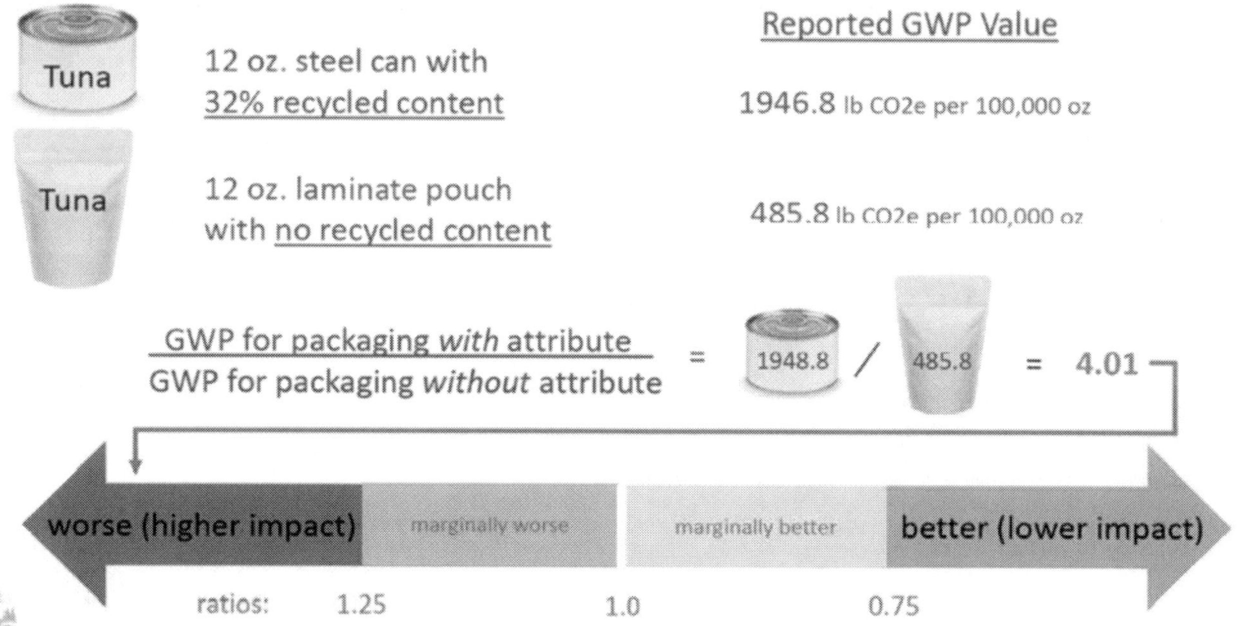

method – evaluation framework

Ratio = Impact result with attribute A ÷ Impact result without attribute A		
Category	**Ratio**	**Interpretation**
Meaningfully Lower Life Cycle Impact	<0.75	Suggests the attribute is potentially a good indicator of environmental performance
Marginally Lower Life Cycle Impact	≥0.75 and <1.0	Marginal difference
No difference	1.0	No difference
Marginally Higher Life Cycle Impact	>1.0 and ≤1.25	Marginal difference
Meaningfully Higher Life Cycle Impact	>1.25	Attribute is potentially not a good indicator of environmental performance

The lower the ratio value, the lower the environmental impact of the material(s) being evaluated (*with* the attribute) compared to the equivalent material *without* the attribute.

visualized in Likert charts

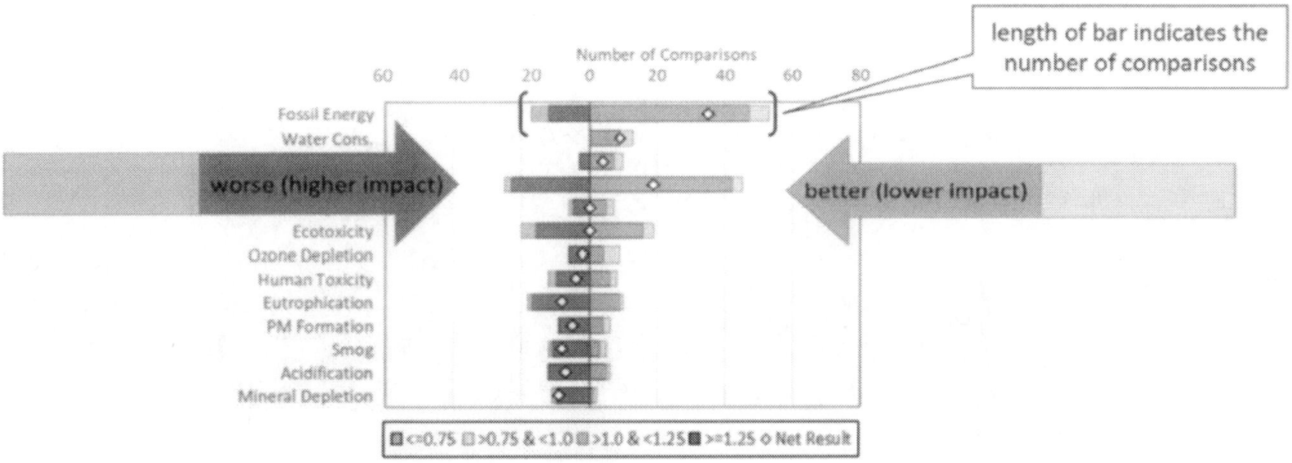

Minal Mistry | Oregon Department of Environmental Quality 7-Jan-20 21

recycled content – packaging

The portion of materials used in a product that have been diverted from the solid waste stream.

Minal Mistry | Oregon Department of Environmental Quality 7-Jan-20 22

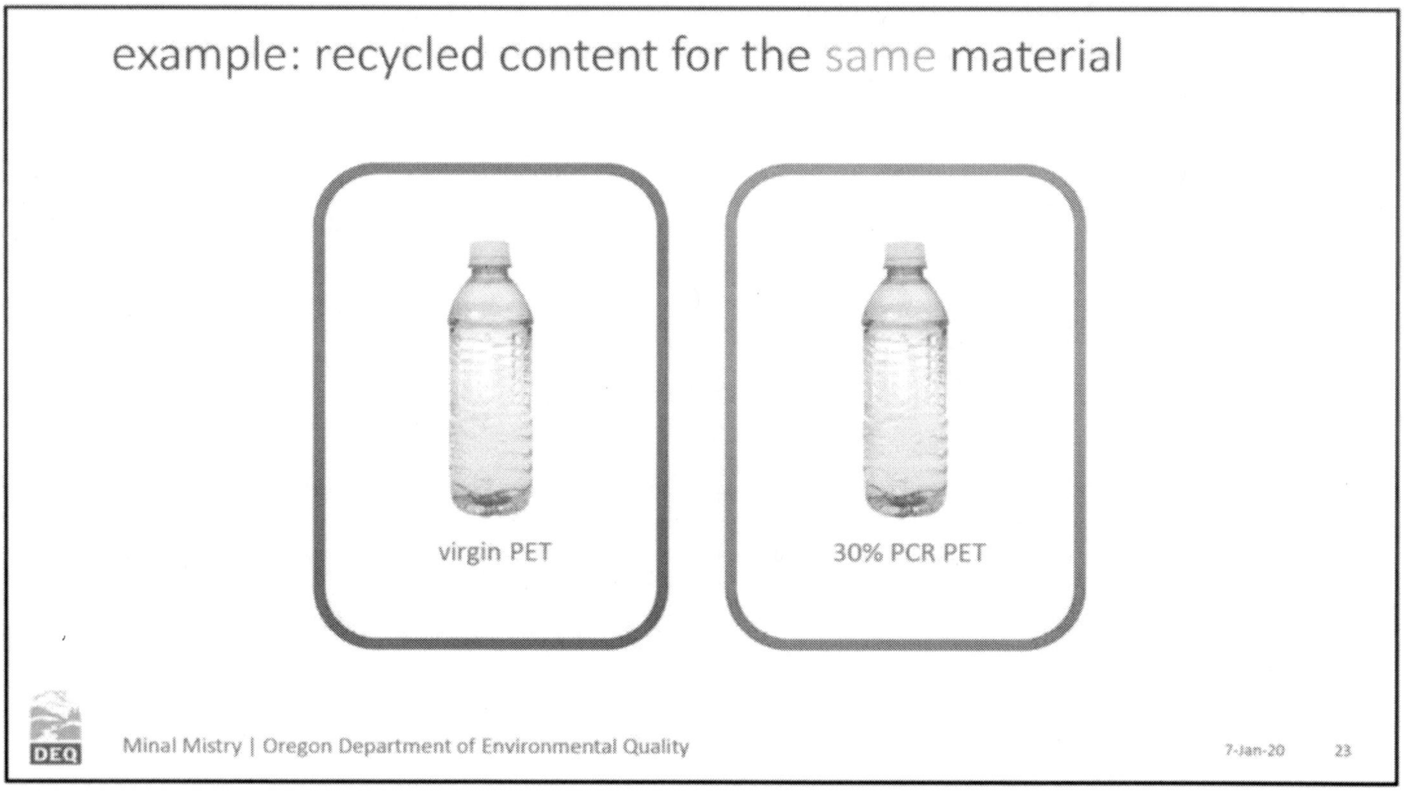
example: recycled content for the same material
virgin PET
30% PCR PET
DEQ
Minal Mistry | Oregon Department of Environmental Quality
7-Jan-20
23

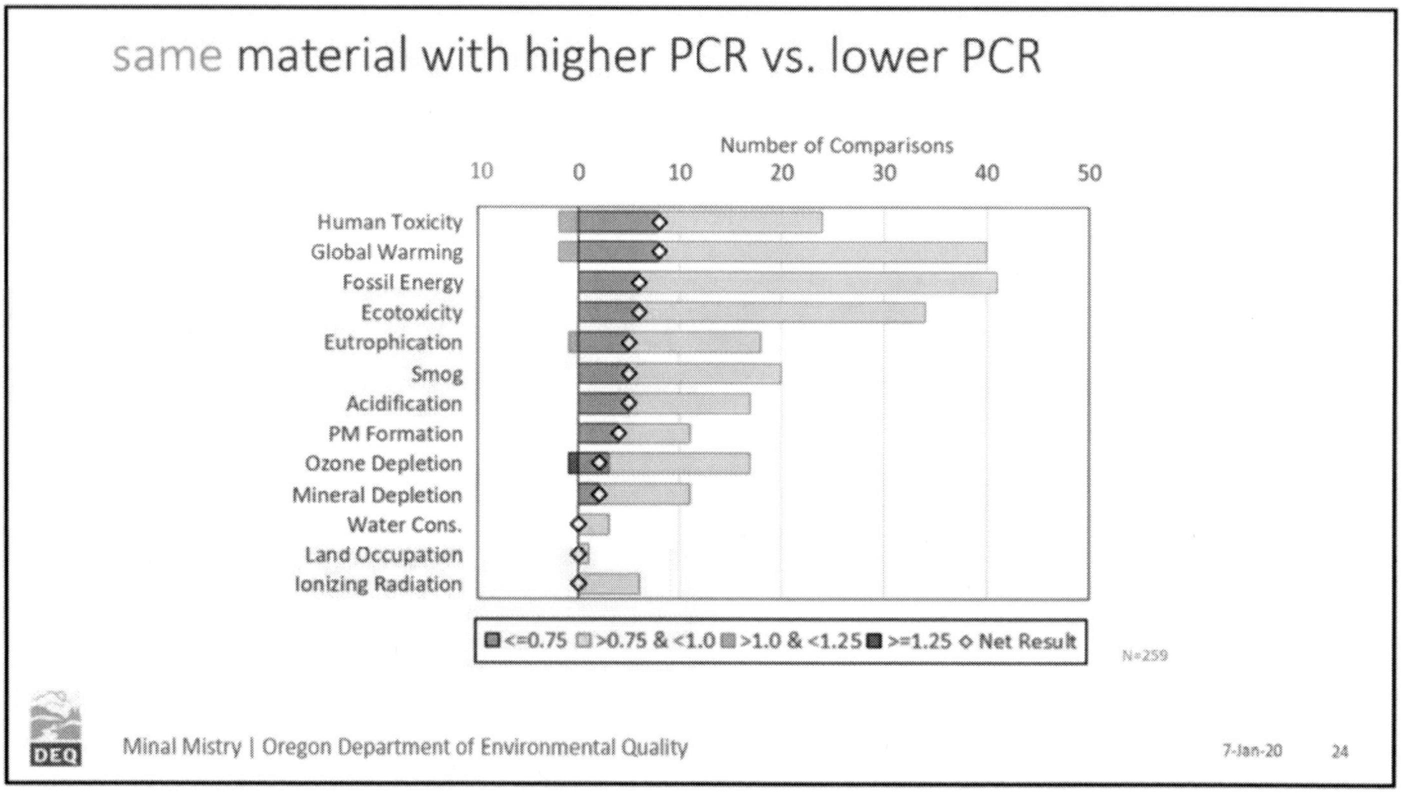
same material with higher PCR vs. lower PCR
Number of Comparisons
10
0
10
20
30
40
50
Human Toxicity
Global Warming
Fossil Energy
Ecotoxicity
Eutrophication
Smog
Acidification
PM Formation
Ozone Depletion
Mineral Depletion
Water Cons.
Land Occupation
Ionizing Radiation
<=0.75 >0.75 & <1.0 >1.0 & <1.25 >=1.25 Net Result
N=259
DEQ
Minal Mistry | Oregon Department of Environmental Quality
7-Jan-20
24

example: recycled content across different materials

steel container
with recycled content

laminate container
without recycled content

DEQ Minal Mistry | Oregon Department of Environmental Quality 7-Jan-20 25

comparing different materials based on PCR (100/0)

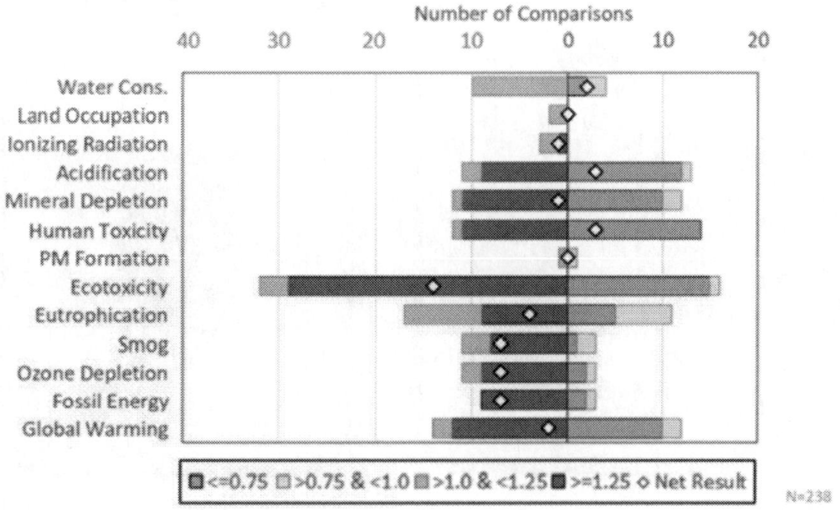

DEQ Minal Mistry | Oregon Department of Environmental Quality 7-Jan-20 26

recyclable packaging

Note: recyclable {the design or purchasing attribute} ≠ recycling {the end-of-life treatment for materials}

DEQ Minal Mistry | Oregon Department of Environmental Quality 7-Jan-20 27

example: recyclable packages of different materials

glass container
that is recyclable

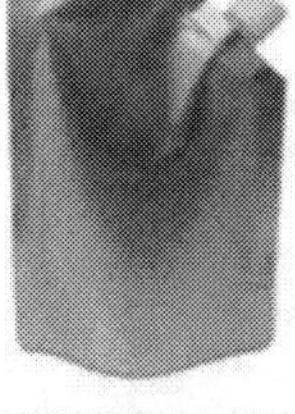

laminated container
that is not recyclable

DEQ Minal Mistry | Oregon Department of Environmental Quality 7-Jan-20 28

comparing different materials based on recyclability (0/100)

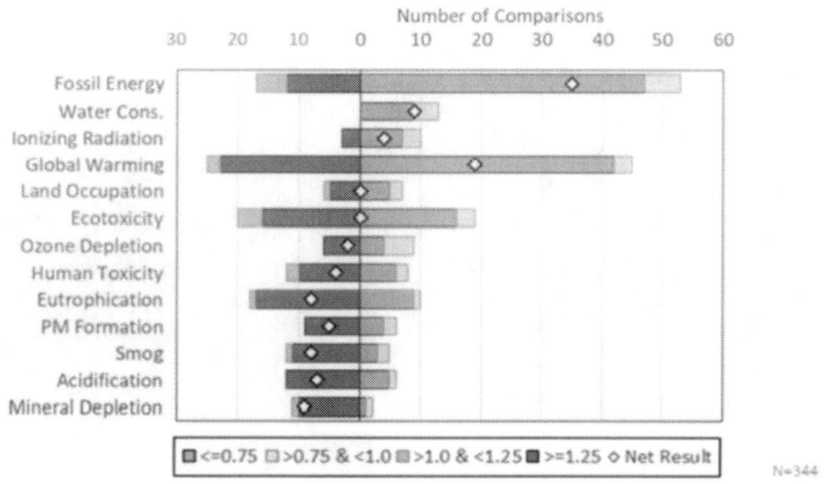

Minal Mistry | Oregon Department of Environmental Quality 7-Jan-20 29

Do popular attributes correlate (reliably) with reduced environmental impacts?

Recycled content (same material)	Yes
Recycled content (different material)	No
Recyclable	No
Biobased	No
Compostable	No

Minal Mistry | Oregon Department of Environmental Quality 7-Jan-20 31

Takeaway #2

Popular packaging attributes do not consistently identify options with lower environmental impacts.

time to expand our portfolio

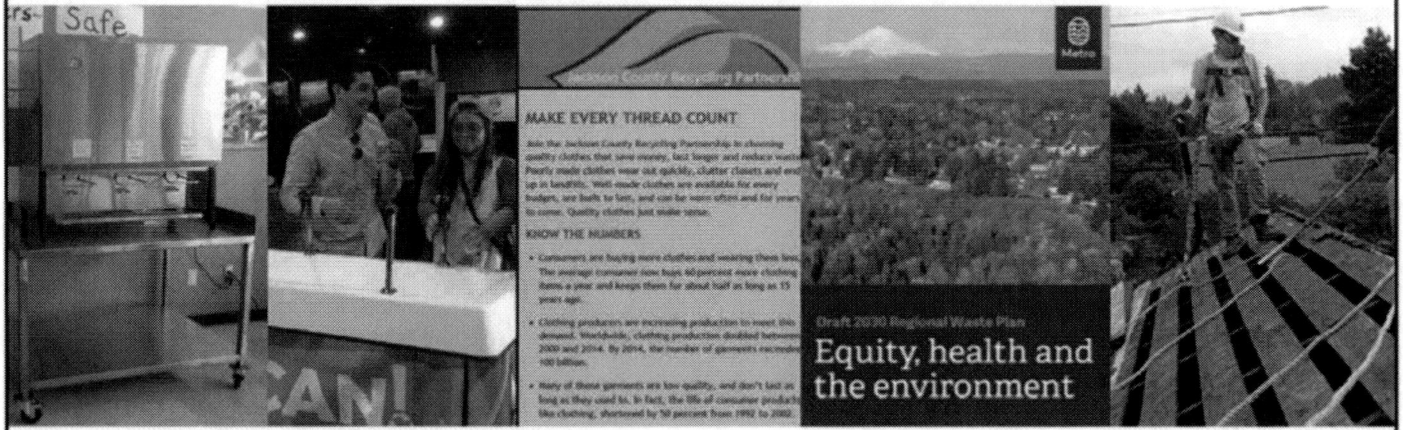

Clackamas County milk dispenser program in schools reduces waste: packaging and waste of milk.

City of Eugene is moving up the waste hierarchy and to reuse with water bottle filling stations at public events.

Jackson County Recycling Partnership is promoting a "Make Every Thread Count" to promote maintenance, repair and reuse of clothes.

Metro's newly adopted Regional Waste Plan aligns with DEQ's 2050 Vision and takes a more holistic life cycle view of materials.

The City of Portland has a deconstruction program to reclaim materials from old houses creating new jobs and markets for aged lumber.

time to expand our portfolio

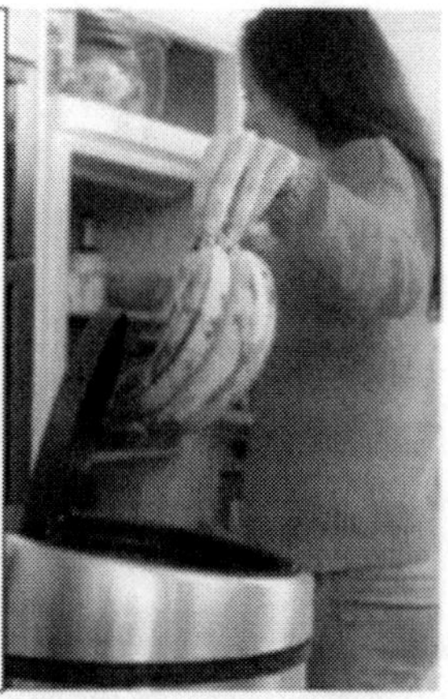

SAFER CHOICE
Meets U.S. EPA Safer Product Standards
epa.gov/saferchoice

Refurbishment and repair "fix-it" fairs, held by the City of Eugene.

Oregon's Concrete EPD program provides tools, technical support, and reimbursement to producers in Oregon.

Increase supply and demand for smaller housing options in Oregon.

Statewide campaign to prevent the wasting of food supported by specific grant funding.

Statewide purchasing agreement to support state spend towards toxics reduction in cleaning products.

DEQ Minal Mistry | Oregon Department of Environmental Quality 7-Jan-20 34

we can make some crucial shifts now

a. Prioritize environmental outcome and source reduction over diversion
b. Procure using a balanced approach to reduce impacts
c. Demand design optimization for life cycle impacts (including, not exclusively for, EoL outcomes)
d. Demand producer responsibility!
e. Recycle smartly – based on environmental benefit and viable secondary market (at times disposal may be the best option)
f. Implement & support policies that reduce environmental burdens and EoL fates (not just via recycling or composting)

DEQ Minal Mistry | Oregon Department of Environmental Quality 7-Jan-20 35

materials management

conserving resources · protecting the environment · living well

minal mistry | minal.mistry@state.or.us

AS YOU SOW

AS YOU SOW

Shareholder Engagement on Plastic Pollution

Sustainable Packaging Symposium

Dec. 17, 2019

1

AS YOU SOW

Shareholder Advocacy

Mobilizing Financial Markets for Social Change

Corporate Engagement

Shareholder Proposals

Proxy Voting

2

AS YOU SOW

8 million tons of plastic waste enter oceans annually

3

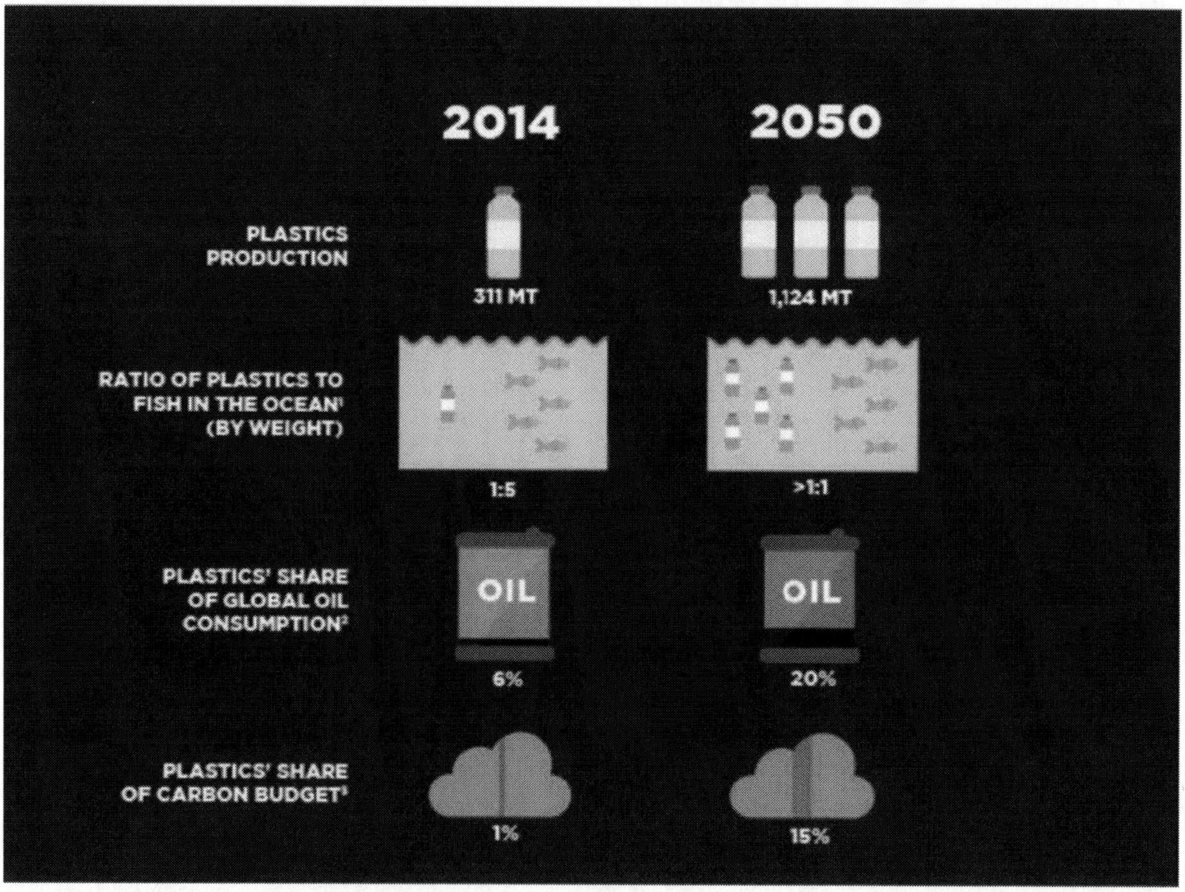

AS YOU SOW

Top 10 Items Collected

Ocean Conservancy

1. CIGARETTE BUTTS 2,127,565
2. PLASTIC BEVERAGE BOTTLES 1,024,470
3. FOOD WRAPPERS 888,589
4. PLASTIC BOTTLE CAPS 861,340
5. STRAWS, STIRRERS 439,571
6. OTHER PLASTIC BAGS 424,934
7. GLASS BEVERAGE BOTTLES 402,375
8. PLASTIC GROCERY BAGS 402,122
9. METAL BOTTLE CAPS 381,669
10. PLASTIC LIDS 351,585

5

AS YOU SOW

Recent studies on environmental impact of plastic

American Chemistry Council

Environmental cost of plastic use $139 billion/year

TRUCOST

UNEP

Plastic packaging causes $40 billion in damage, including $13 billion damage to marine ecosystems per year

6

AS YOU SOW

Wasted Value

- **95% of plastic packaging value, $80–120 billion, lost to economy after short first use.**
- **No closed loop:** Most plastics that are recycled converted to lower-value applications that represent final use.
- **Open Loop:** 80% of recycled PET bottles go to polyester fibers for carpet, clothing. Other plastics go to 'plastic lumber', pipes, and waste collection bags.

7

AS YOU SOW

Multi-layer Flexible Pouches Moving to Dominate Market

8

AS YOU SOW

Flexibles are dominating packaging

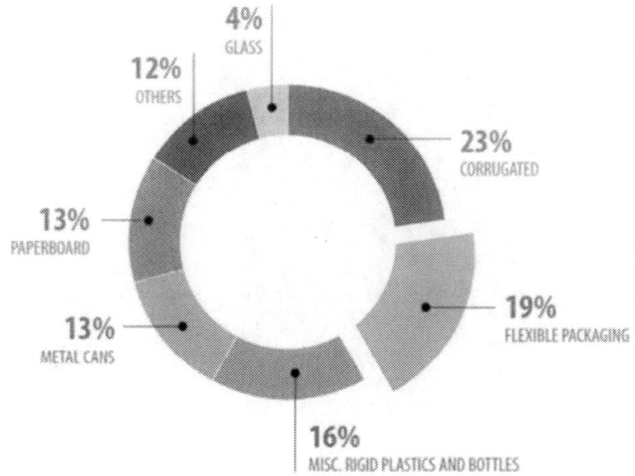

- $34 billion in annual sales in U.S. sales
- Boosts shelf visibility
- Lightweight
- Low GHG production profile...
- ...But Not recyclable

9

AS YOU SOW

Steel versus pouch

10

AS YOU SOW

Cardboard box versus pouch

11

AS YOU SOW

Bottle versus pouch

12

AS YOU SOW

Potential for Brand Risk

13

AS YOU SOW

Coca-Cola, PepsiCo, and Nestlé found to be worst plastic polluters worldwide in global cleanups and brand audits

by Jed Alegado | Oct 9, 2018 | Brand Audit, Featured, Press Release | 11 comments

Recycling Commitments to As You Sow

pepsi

2010 - 50% beverage industry bottle/can recycling rate by 2018
Result – 36%

Nestlé Waters

2009 - 60% PET plastic bottle recycling rate by 2018
Result – 30%

Global policy developments

200 nations pledge to eliminate plastic pollution from oceans at Nairobi UN Environment Assembly

- 50+ countries banned plastic bags
- Nine countries banned Styrofoam packaging
- U.S. microbead ban in 2017
- UK, Scotland enacting container deposit laws
- EC moving to ban some single use plastics

AS YOU SOW

As You Sow
Engagement on plastic pollution

Agenda

- Make all packaging recyclable
- Corporate responsibility to finance recycling
- Priority on polystyrene foam phase out
- Redesign of non-recyclable pouches, sachets
- Transition to refillables, reusables
- Plastics use transparency and reduction

17

AS YOU SOW

Colgate-Palmolive commitment

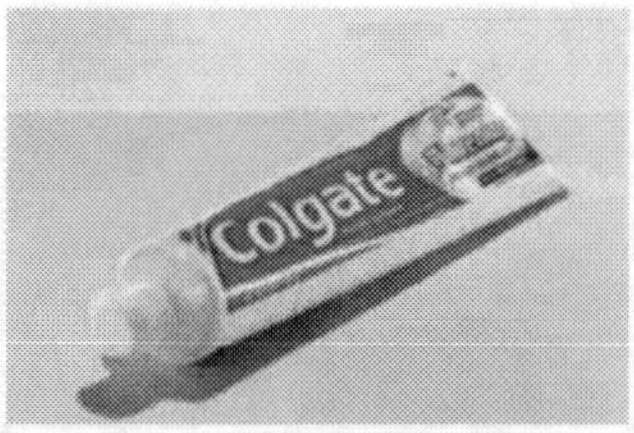

- Make packaging recyclable in 3 product categories by 2020
- Recyclable toothpaste tube

18

P&G Commitment

By 2020, make 90% of packaging recyclable, or “programs are in place to create ability to recycle it”

Recyclable Packaging Commitments

Mondelēz International

KraftHeinz

Unilever

AS YOU SOW

McDonald's PS foam phase out
2017 shareholder vote 32%
$25 billion share value

21

AS YOU SOW

Starbucks 2018 vote – 29%
$54 billion in share value

Ban straws

Redouble efforts on reusables

More recyclable cups

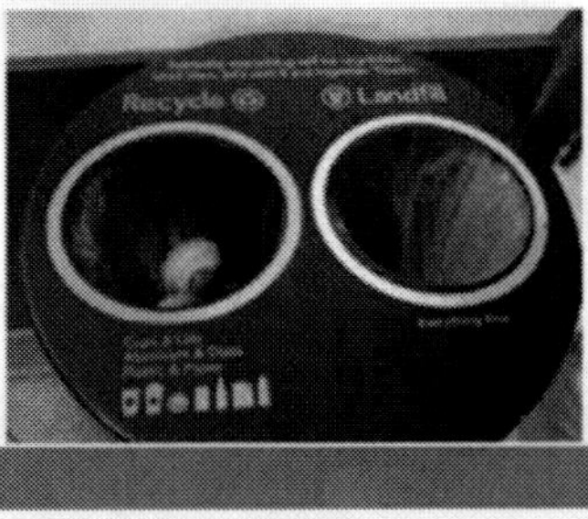

22

AS YOU SOW

Plastic Solutions INVESTOR ALLIANCE

40 investors
$2.5 trillion in collective assets under management

Dialogues with large consumer goods companies

Ask:
- Make it recyclable, reusable, or compostable
- Disclose annual plastic packaging use
- Set plastic use reduction goals
- Develop alternatives to single use packaging
- Accept producer responsibility for recycling
- Support public policy measures to reduce plastic waste

23

AS YOU SOW

Recent Corporate Plastics Reduction Commitments

- **Unilever** - Cut plastics use by 100,000 tons and cut use of virgin plastic by half (2025)
- **Pepsi** - Reduce use of virgin plastic 35% on beverage side by 2025, potential reduction of virgin plastic use of 2 million tons
- **Nestle**: *"We are determined to reduce our use of single-use plastics..."*

24

AS YOU SOW

Proposed Solutions

- Systems Focus: Companies need to look beyond their own narrow product impact
- U.S. National platform for joint stakeholder engagement
- Design packaging for recyclability
- Producer Responsibility

25

AS YOU SOW

Proposed Solutions

- Corporate culture change – Packaging designers, marketers, sustainability staff, and recyclers need to be aligned
- Reduced single use plastic applications through alternative delivery systems
- Build domestic recycling markets

26

AS YOU SOW

Contact

Conrad MacKerron

As You Sow

mack@asyousow.org

Thank you!

27

Liviri

Reusable solutions for e-Commerce perishables and last-mile delivery

PRODUCT QUALITY FOR TEMP-SENSITIVE GOODS

SUPERIOR OPERATIONAL EFFICIENCIES

REDUCING ENVIRONMENTAL FOOTPRINTS

Liviri

LIVIRI PORTFOLIO

Rugged, high performance, shipping coolers, designed to ship in SMALL PARCEL NETWORK

Liviri *fresh*

Liviri *vino*

Portfolio of durable, less expensive LAST MILE DELIVERY CONTAINERS

Liviri *shuttle*

Liviri

SUSTAINABILITY APPROACH

1. Engineer rugged products
2. Design for recyclability & disassembly
3. Implement closed loop system

Liviri

Liviri Fresh – Life Cycle Assessment

- **Otter Products Liviri Fresh**
  - Reusable vacuum insulated panel (VIP) cooler
- **Corrugate cooler (NA average)**
  - Single use, corrugated box cooler with recycled cotton insulation

VS.

Liviri

Liviri Fresh – Life Cycle Assessment

- **Function:** Providing transport packaging for three two-person meals
- **Functional unit:** Providing transport packaging for three two-person meals for 1 year
- **Lifetime:** Corrugated cooler: 1 delivery
  Liviri Fresh: 50 deliveries (conservative)
- **Reference flows:** (365/3) = 121.7 corrugate coolers
  (365/3/50) = 2.43 Liviri Fresh coolers

Liviri

Liviri Fresh – Life Cycle Assessment

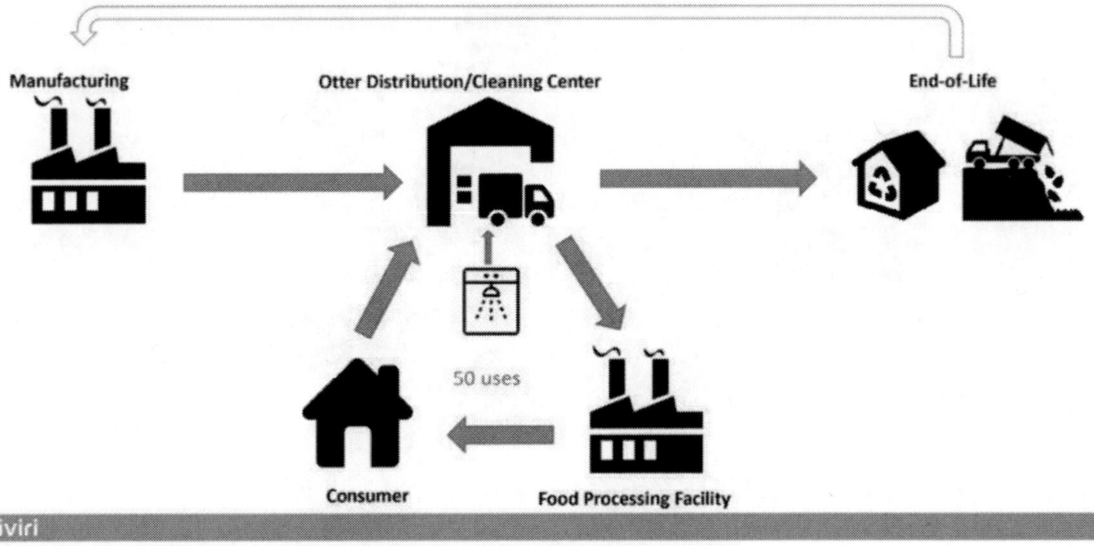

Liviri

Liviri Fresh – Life Cycle Assessment

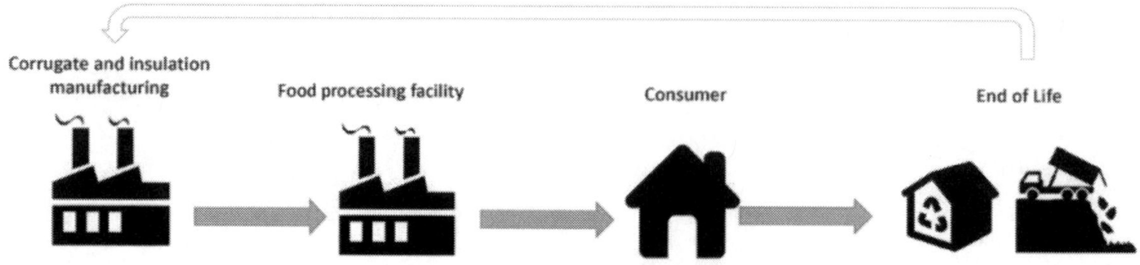

Single Make-Use-Dispose Life Cycle

Liviri

Liviri Fresh – Life Cycle Assessment

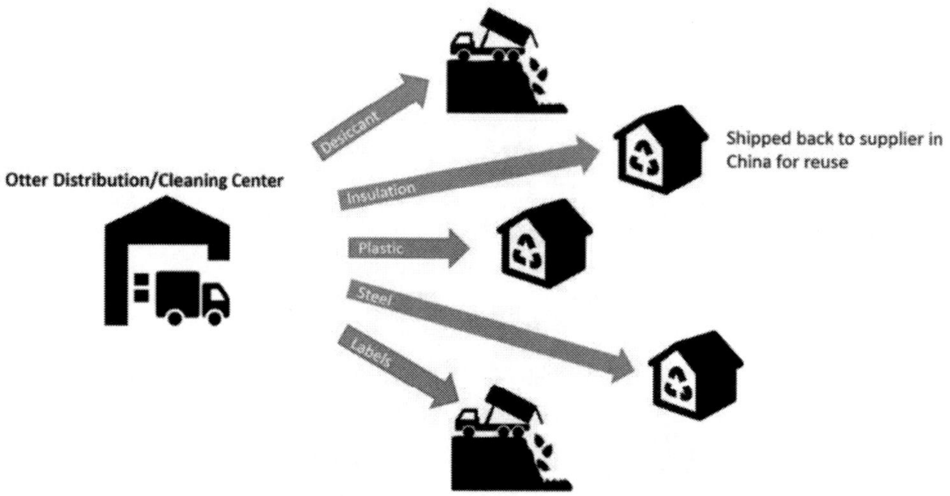

Liviri

Liviri Fresh – Life Cycle Assessment

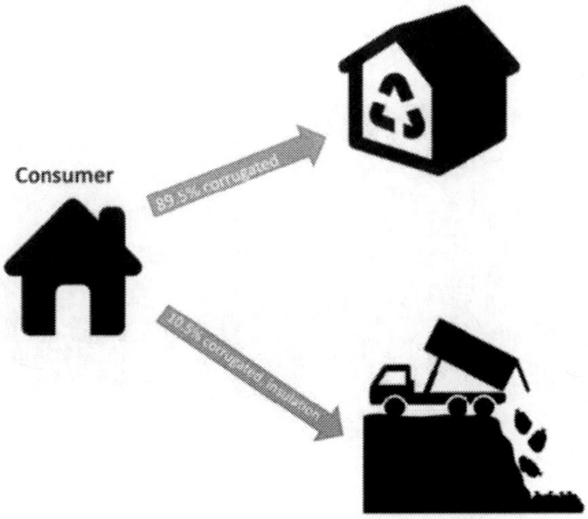

Liviri Fresh – Life Cycle Assessment

Life Cycle Impact Assessment Categories

- Global Warming (GWP100 fossil) [kg CO2-equiv.] excluding biogenic CO2
- Global Warming (GWP100 total) [kg CO2-equiv.] including biogenic CO2
- Acidification (AP) [kg SO2-equiv.]
- Eutrophication (EP) [kg Phosphate-equiv.]
- Particulate Matter (PM) [kg PM2.5-equiv.]
- Smog Formation Potential (SFP) [kg O3-equiv.]

Life Cycle Inventory Metrics

- Primary Energy Demand (PED) [MJ NCV]
- Blue Water Consumption (BWC) [m3]

Liviri Fresh – Life Cycle Assessment

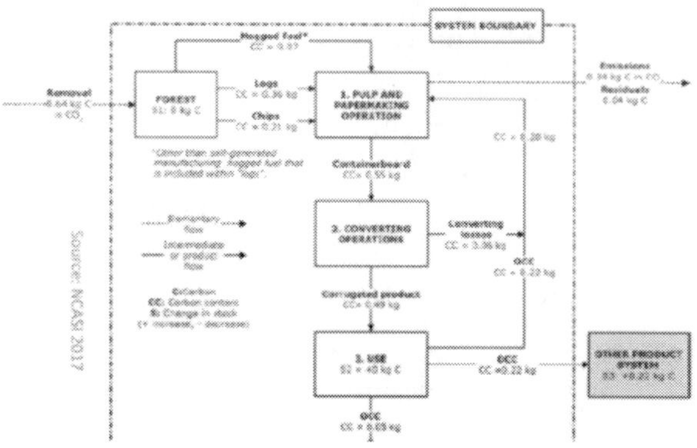

CPA Carbon Flow Accounting

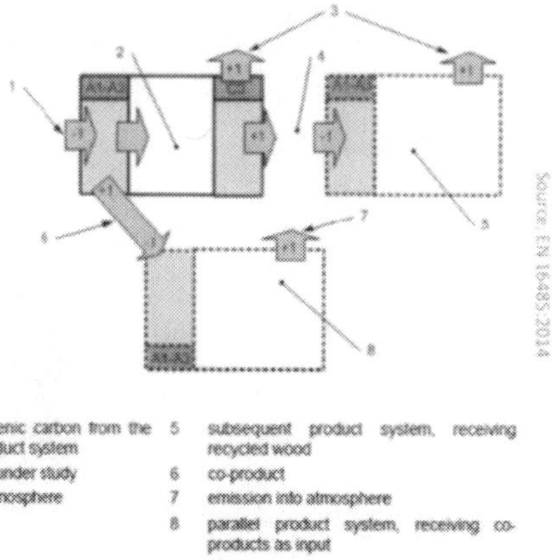

Full Carbon Accounting (baseline)

Liviri

Liviri Fresh – Life Cycle Assessment

Liviri Fresh has lower impact category and inventory indicator results than the disposable corrugated alternative in 8 of the 9 indicators considered.

- Recycled cotton insulation and some corrugated product landfilled after one use; not 'circular' BUT...
- Significant share of the biogenic C becomes permanently sequestered in the landfill

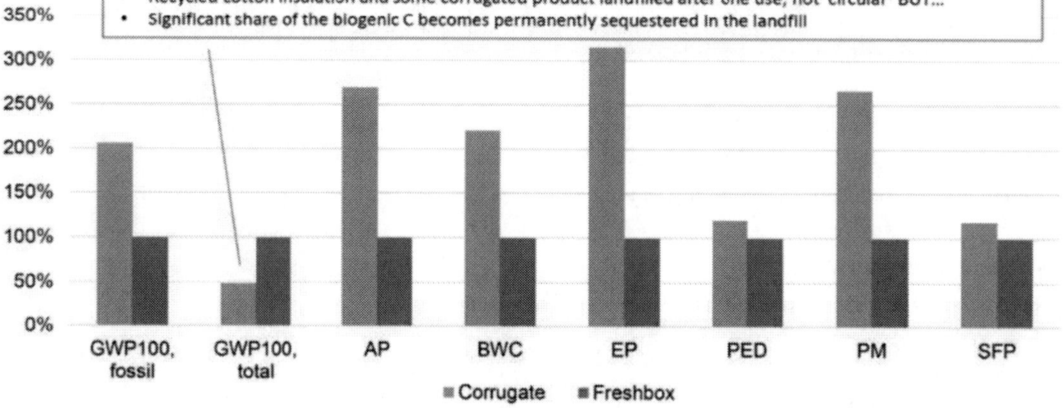

Liviri

Liviri Fresh – Life Cycle Assessment

Liviri Fresh

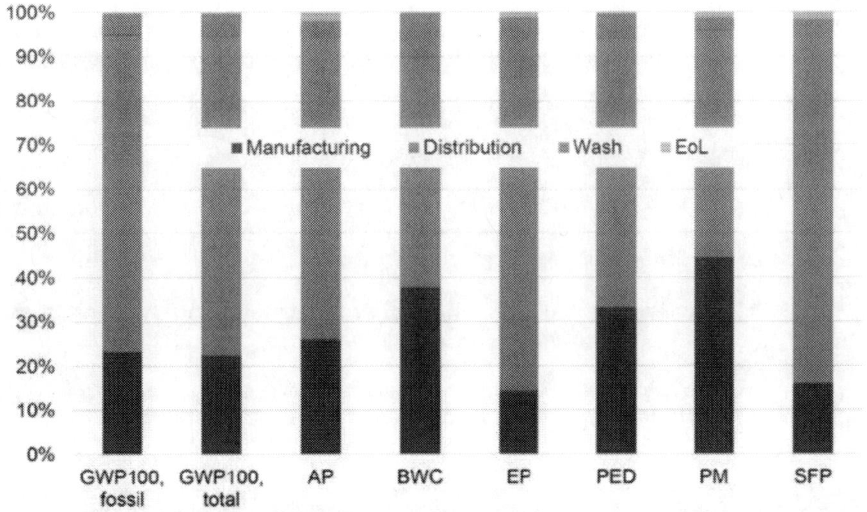

Liviri

Liviri Fresh – Life Cycle Assessment

Single-use

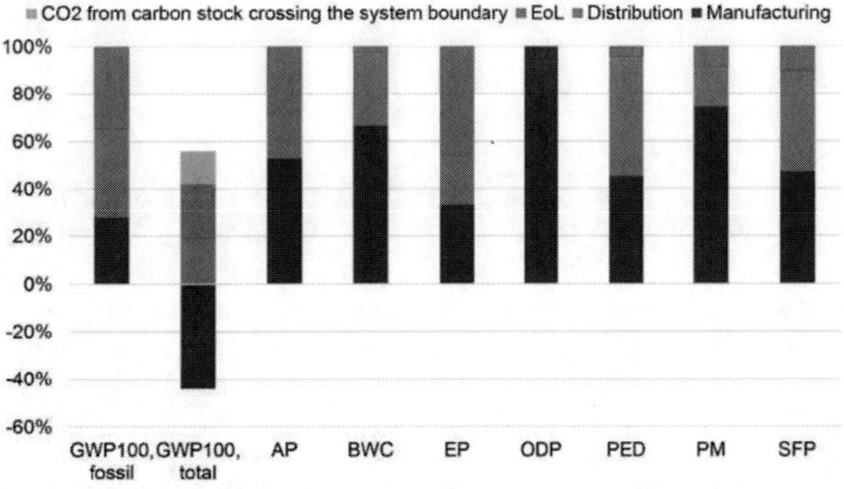

Liviri

Liviri Fresh – Life Cycle Assessment

Product	Full Carbon Accounting	CPA Carbon Accounting
Corrugated Cooler	110	423
Liviri Fresh, 50 uses	226	219

- CPA's own approach leads to a higher total GWP100 result for the corrugated product since the carbon contained in the recycled cotton insulation is not associated with any carbon removals.
- Accordingly, the effect of permanently sequestrating the contained carbon in the landfill is eliminated from the inventory.
- Since total GWP100 results hence depend on a *normative modeling choice*, only GWP100 fossil result were considered to be robust enough to inform final conclusion.

Liviri

Liviri Fresh – Life Cycle Assessment

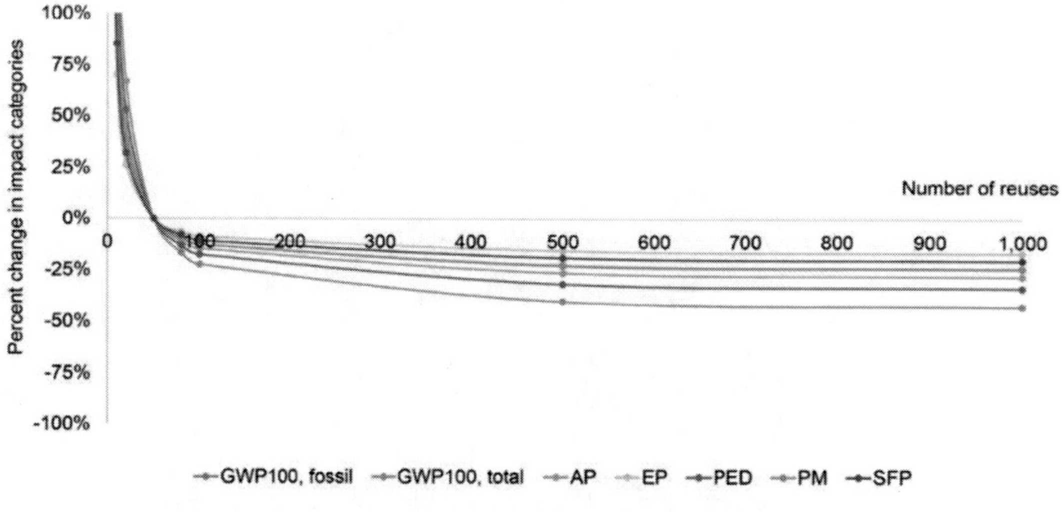

Liviri

Liviri Fresh – Life Cycle Assessment

In conclusion…

- Due to its high reusability, the Liviri Fresh shows better environmental performance than the single-use corrugated design

- Optimization potentials identified for the Liviri Fresh included:
  - Maximizing lifetime (>>100 uses possible, 50 assumed in LCA)
  - Reducing air-freight upstream for several parts
  - Low-emission trucking services (CNG, advanced biofuels, electric)

Liviri

Liviri Vino – Life Cycle Assessment

Liviri Vino
- Reusable vacuum insulated panel (VIP) cooler
- ~~Reusable ice pack~~
- Unrefrigerated ground shipments

Single-use cooler
- Single-use, EPS & corrugate cooler
- ~~Single-use ice pack~~
- 2-day air shipments

VS.

Liviri

Liviri Vino – Life Cycle Assessment

- Liviri Fresh LCA (and packaging LCAs in general) only consider the weight of the packaging itself to calculate burdens associated with transportation throughout the life cycle
- Fine *ONLY IF* the transportation modes are *NOT* dependent on the packaging design
- For Liviri Vino, packaging eliminates the need for air freight *BUT* single-use packaging still light enough to compensate the difference
- The full benefits of eliminating air freight can only be accounted for if weight of the wine and wine bottle is also considered
- Need to include wine and bottles from cradle-to-grave → ~~packaging~~ LCA of wine consumption

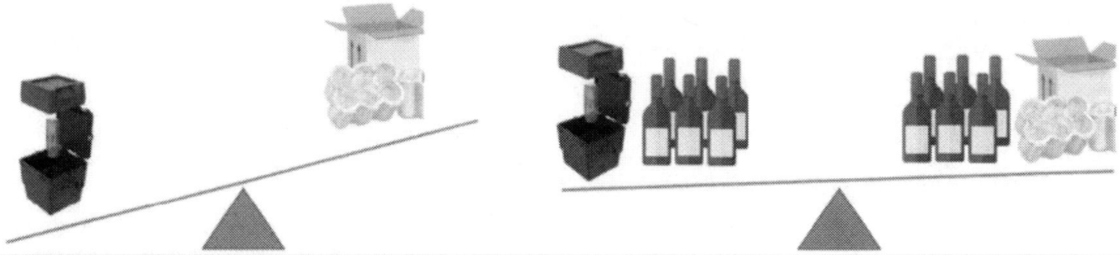

Liviri

Liviri Vino – Life Cycle Assessment

- **Function:** Providing transport packaging shipments of 6 bottles of wine
- **Functional unit:** Provision of 6 bottles of wine and transport packaging delivered seventy times over a two-year period (2*365/10 ≈ 70 deliveries)
- **Lifetime:** Liviri Vino = 70 deliveries
Single-use cooler = 1 delivery
- **Reference flows:** 70 single use coolers
1 Liviri Vino

Liviri

Liviri Vino – Life Cycle Assessment

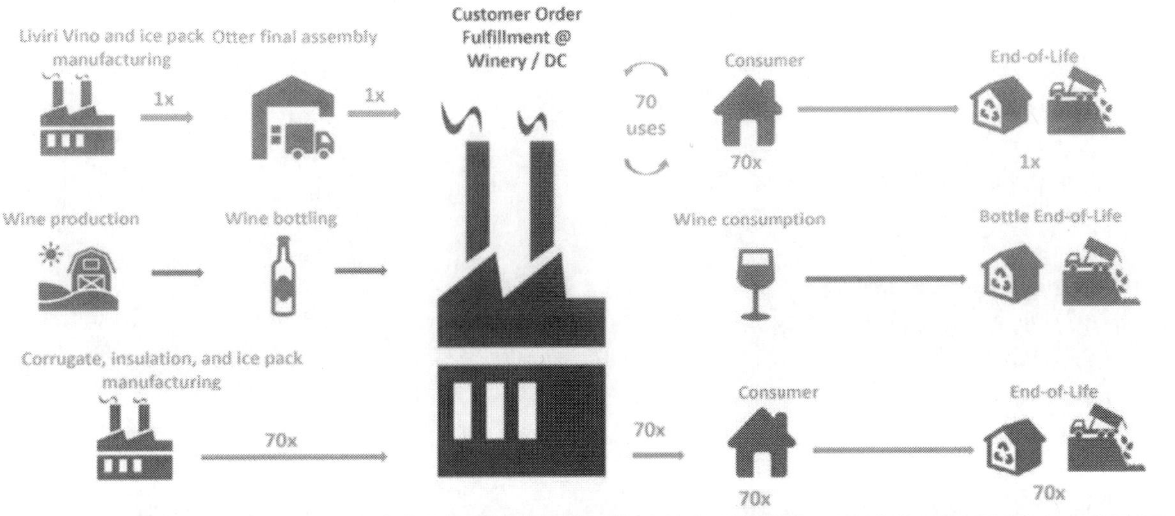

Liviri

Liviri Vino – Life Cycle Assessment

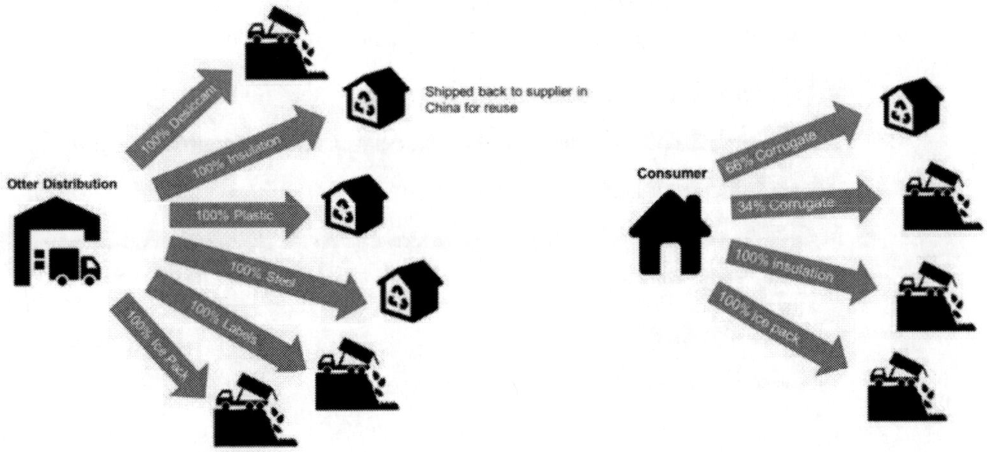

Liviri

Liviri Vino – Life Cycle Assessment

Life Cycle Impact Assessment Categories

- Global Warming (GWP100 fossil) [kg CO2-equiv.] excluding biogenic CO2
- Global Warming (GWP100 total) [kg CO2-equiv.] including biogenic CO2
- Acidification (AP) [kg SO2-equiv.]
- Eutrophication (EP) [kg Phosphate-equiv.]
- Particulate Matter (PM) [kg PM2.5-equiv.]
- Smog Formation Potential (SFP) [kg O3-equiv.]

Life Cycle Inventory Metrics

- Primary Energy Demand (PED) [MJ NCV]
- Blue Water Consumption (BWC) [m3]

Liviri

Liviri Vino – Life Cycle Assessment

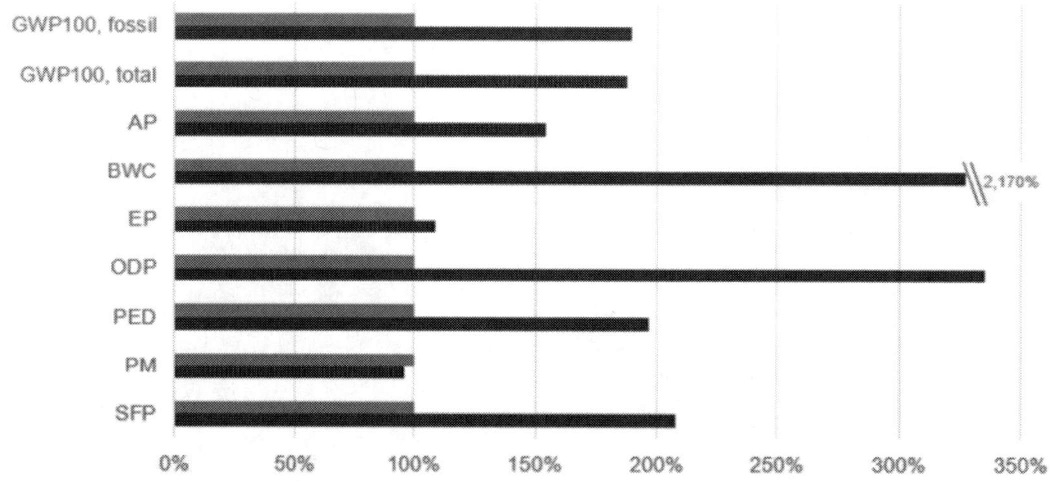

Liviri

Liviri Vino – Life Cycle Assessment

Liviri Vino

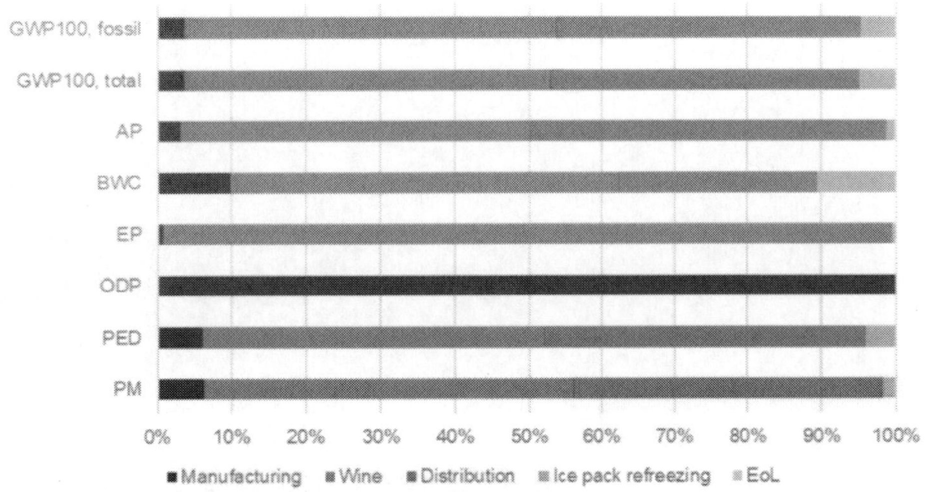

Liviri

Liviri Vino – Life Cycle Assessment

Single-use

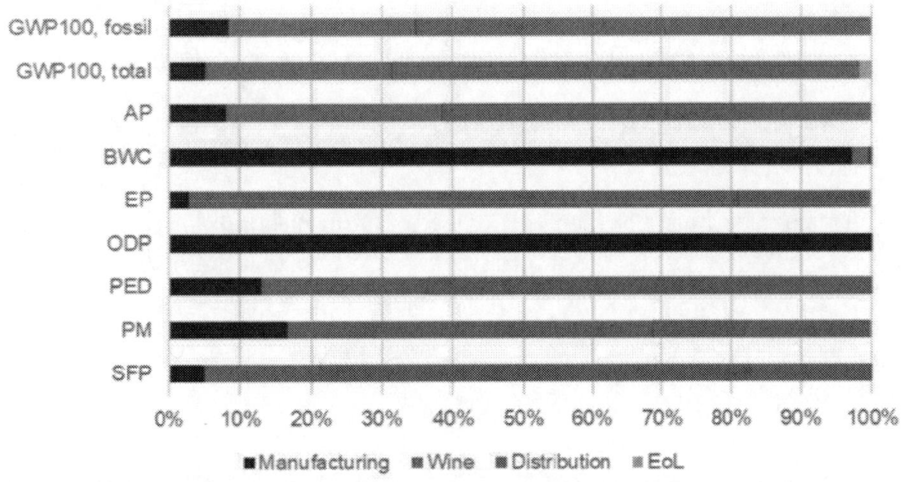

Liviri

Liviri Vino – Life Cycle Assessment

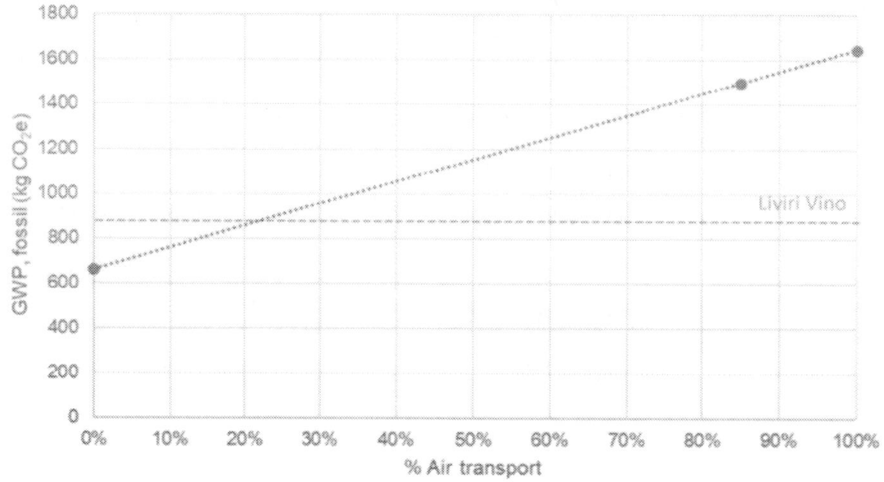

Liviri

Liviri Vino – Life Cycle Assessment

In conclusion...

- Due to its high reusability and eliminating the necessity of air freight, the Liviri Vino shows better environmental performance than the single-use corrugated design
- If the packaging design is causally responsible for such a shift in transportation mode, it is necessary to account for the total change in ton-miles ***including the packaged product***
- Optimization potentials identified for the Liviri fresh included:
  - Maximizing lifetime (>>100 uses possible, 70 assumed in LCA)
  - Low-emission trucking services (CNG, advanced biofuels, electric)

Liviri

Lesson Learned about Sustainable Packaging

Looks can be deceiving!

- Cardboard not automatically better than plastic just because it is bio-based and has a decent recycled content
- "Single-use" associated with plastics, but using *any material* just once is a problem.

It's the performance, stupid!

- VIP provide superior thermal performance which reduces demand for ice (shipped mass, energy) and allow for longer shipping times which can help reduce air freight

It's the longevity, stupid!

- Despite higher transportation demands of reuse cycles and additional cleaning, keeping packaging products in use for as long as possible can pay off

It's the business model, stupid!

- Liviri offers a lease model where they retain ownership of the packcging
- Broken units get recycled into new Liviri products

Lesson Learned about Sustainable Packaging

Product perspective beats material perspective

- No material is "bad" per se (notwithstanding immediate health hazards etc.)
- Plastic packaging can perform very well from an environmental perspective if used the right way

Life is complicated...

- None of this can be generalized beyond Liviri
- A similar study on reusable plastic shipping containers did not lead to such clear-cut results, BUT...
- Food losses can really tip the scale in favor of more durable products
- Case-by-case assessments necessary

Final Thoughts

Beware of "analysis paralysis"

- Attributional LCA does not consider consequences on other product systems or society as a whole
- Data is mostly generic or average, large and mostly unknown uncertainties remain
- Economic modeling far more uncertain and complex than mass and energy based models
- You can only ever know so much; don't let the perfect be the enemy of the good
- Evolution is trial-and-error; make a reasonable effort to identify and avoid unintended consequences once, but deal with them as they occur (or not)

A Different View of Source Reduction

Conference: Sustainable Packaging Symposium (SPS)

Year: 2019

Proceeding: Sustainable Packaging Symposium 2019

Group: Abstract Submissions

Session: Panel III – Measurable Progress in Packaging Sustainability - Part 2

Time: Tuesday, December 17, 2019 - 1:50pm-2:15pm

Authors:

Wagger, D., Institute of Scrap Recycling Industries (ISRI)

According to EPA, Source Reduction â€œis any practice that reduces, eliminates, or prevents pollution at its sourceâ€. It resides at the top of EPAâ€™s Environmental Protection Hierarchy, followed by Re-Use, Recycling, Treatment, and Disposal (Release), from more to less preferable. As defined in the Pollution Prevention Act of 1990, Source Reduction â€œmeans any practice which ... reduces the amount of any hazardous substance, pollutant, or contaminant entering any waste stream or otherwise released into the environment (including fugitive emissions) prior to recycling, treatment, or disposalâ€. It further â€œincludes equipment or technology modifications, process or procedure modifications, reformulation or redesign of products, substitution of raw materials, and improvements in housekeeping, maintenance, training, or inventory controlâ€. Both the PPA and EPA acknowledge existing regulatory barriers to Source Reduction opportunities. A common strategy seems to be mass reduction of packaging and products, in part to reduce energy use and emissions during manufacturing, transport and/or use. Such dematerialization and other Source Reduction strategies affect Re-Use, Recycling, Treatment, and Disposal (Release) differently. The presentation will explore some of these differences, especially potential challenges and opportunities for Recycling.

Using Streamlined LCA to Assess Packaging Requirements in Aerospace and Defense

Conference: Sustainable Packaging Symposium (SPS)

Year: 2019

Proceeding: Sustainable Packaging Symposium 2019

Group: Abstract Submissions

Session: Panel III – Measurable Progress in Packaging Sustainability - Part 2

Time: Tuesday, December 17, 2019 - 2:15pm-2:40pm

Authors:

Richard, C., Lockheed Martin

At Lockheed Martin, we use a variety of streamlined life cycle assessment (S-LCA) tools to provide greater insight and understanding of our environmental impacts. Our approach to S-LCA is based directly on the Dept. of Defense's Guidance on Sustainability Analysis is a streamlined method for assessing life cycle impacts (LCA) and cost (LCC). Through this streamlined approach we can quickly and efficiently conduct comparative assessments and return value to our business and supply chain partners. We have also found that such streamlined approaches to life cycle thinking complement traditional Cost/Benefit analyses, while providing a greater emphasis on the social and environment value. This presentation will provide an overview of the methods used and summarize an example of how we assessed packaging options related to an industrial scale product.

Life Cycle Environmental Impact Assessment of Epoxidized Sucrose Soyate: A Soy-Based Adhesive

Conference: Sustainable Packaging Symposium (SPS)

Year: 2019

Proceeding: Sustainable Packaging Symposium 2019

Group: Abstract Submissions

Session: Technical Session 2 Research in Packaging Sustainability

Time: Tuesday, December 17, 2019 - 12:00am-12:25am

Authors:

Pourhashem, G., North Dakota State University
Ghasemi, S., North Dakota State University
Cecchin, A., North Dakota State University
Webster, D., North Dakota State University

In the past few decades, the demand for renewable and sustainable polymeric materials has significantly grown due to 1) their ability to provide enhanced material functionality, and 2) increasing environmental and health concerns from fossil-based products. Epoxidized vegetable oils and their derivatives can provide alternative sources for petroleum-based resins. Among epoxidized vegetable oil-based resins, epoxidized sucrose soyate (ESS) has shown promises in significantly enhancing the mechanical properties as an adhesive. Though inherently biobased, it is essential to ensure these resins can offer a more sustainable alternative compared to their petroleum-based counterparts.

Our study focuses on investigating the environmental performance of the ESS resins. We use laboratory and pilot-scale data to model the process of preparing ESS resin in industrial-scale and use life cycle assessment (LCA) method to assess the environmental impacts of making this resin. We use SimaPro software and TRACI 2 method (including ten environmental impact categories e.g. global warming) to build our LCA model and compare the environmental performance of ESS with commercially available petroleum-based epoxy resins. Our results show that ESS resins demonstrate a better environmental performance in almost all investigated categories.

Our results will provide decision-making information for its commercialization as well as allow us to identify hotspots in the production process and recommend possible synthesis improvements that can help this novel resin compete with commercially available resins in the market. To our best of knowledge, this is the first research that investigates the environmental impacts of ESS.

LCA of an Innovative Shelf Life Extension Technology

Conference: Sustainable Packaging Symposium (SPS)

Year: 2019

Proceeding: Sustainable Packaging Symposium 2019

Group: Abstract Submissions

Session: Technical Session 2 Research in Packaging Sustainability

Time: Tuesday, December 17, 2019 - 3:20pm-3:45pm

Authors:

Thoits, S., Apeel Sciences
Frazier, C., Apeel Sciences
Vieira, J., Apeel Sciences

Globally one-third to half of all food produced for consumption goes to waste,[1] and fresh fruits and vegetables have the highest food waste rates due to perishability.[2] Many existing solutions that extend produce shelf life, such as refrigeration technologies or single-use plastics, have significant environmental impacts.[3] Apeel Sciences has developed a plant-derived coating for fresh produce that slows the rate of water loss and oxidation, and can keep produce fresh two to three times longer.[4]In October 2019, Apeel completed an LCA study (critical review performed by Quantis[5]) to measure the environmental impacts of incorporating its plant-derived coating into the supply chains for several produce categories. The scope of this study was cradle-to-grave, and the functional unit was 1kg Apeel Produce consumed in a US home. The analysis was conducted in OpenLCA using ecoinvent v3.5 background data and the IMPACT 2002+ LCIA methodology. The results demonstrate that through waste reduction and other supply chain efficiency gains, the lifecycle environmental impacts of Apeel Produce are 6% to 82% lower compared to those of baseline produce consumed in US homes. The presentation will discuss the study methodology and describe how the results can be used to support Apeelâ €™s operations and business practices.

[1] FAO, *Global food losses and food waste â€“ Extent, causes and prevention*, (Rome, FAO, 2011), 4.

[2] FAO, *Global food losses*, 7.

[3] S.J. James & C. James, *The food cold-chain and climate change*, (Food Research International, 43, 2010), 1944-1956.

[4] Apeel website

[5] Quantis website

The Environmental Advantages of Reusable and Insulated Plastic Containers for Shipping Perishable Food Items

Conference: Sustainable Packaging Symposium (SPS)

Year: 2019

Proceeding: Sustainable Packaging Symposium 2019

Group: Abstract Submissions

Session: Technical Session 2 Research in Packaging Sustainability

Time: Tuesday, December 17, 2019 - 3:45pm-4:10pm

Authors:

Koffler, C., thinkstep Inc.
Fanning, K., Otter Products LLC

The landscape of plastic packaging is currently under going public scrutiny. Material substitutions, reductions in demand, investments in recycling capacities and technologies are all part of a more sustainable future. But, the combination of high-performance plastic packaging solutions with circular business models offer the opportunity to leverage plastic's unique material properties while minimizing environmental impact.

Liviri has set out to change the way perishable items are shipped to consumers. At first glance comparing a plastic product against the ubiquitous corrugated cardboard box, which is bio-based, biodegradable, and has a share of recycled content, seems counter-intuitive from a sustainability perspective. However, the high durability, excellent insulative performance, and circular business model of Liviri Fresh and Liviri Vino give some clear advantages over the corrugated incumbent.

To test the environmental performance of their products against competitors, Liviri commissioned thinkstep, a global leader in life cycle assessment and sustainability consulting, to conduct two separate case studies: one on meal kit home delivery (Liviri Fresh) and one on wine home delivery (Liviri Vino). This presentation highlights key differences between Liviri products and their corrugated cardboard competitors in terms of system properties, modeling parameters, and assumptions. We wrap up with a summation of potential environmental impacts for each product system based on an equivalent amount of food items delivered to consumers.

Harmonizing Circular Economy Practices in an ISO Global Standard - Implications for Sustainable Packaging

Conference: Sustainable Packaging Symposium (SPS)

Year: 2019

Proceeding: Sustainable Packaging Symposium 2019

Group: Abstract Submissions

Session: Plenary Session I

Time: Wednesday, December 18, 2019 - 9:15am-9:40am

Authors:

Levy, M., First Environment

Why is standardization important for circular economy (CE) and sustainable packaging?
â€¢ Very different interpretation of what does and does not constitute a â€œcircularâ€ approach

Practical CE applications to date that impact sustainable packaging design (which may inform future standards)
â€¢ May involve design changes, process changes, and restructuring recovery efforts to avoid waste streams, all with the goal of â€œclosing the loopâ€

What the global standards development process for CE should look like
â€¢ Harmonize current CE practices to determine most appropriate definitions and best practices based on stakeholder input

Long-term policy implications of an ISO standard on CE and sustainable packaging
â€¢ A globally accepted definition (and promotion) of circular economy will lead to further policy development and revisions and have an impact on sustainable packaging

Key takeaways that may impact sustainable packaging
â€¢ Participation in (and even just monitoring) standard development will help you remain ahead of the curve in positioning for future best practices
â€¢ Having a seat at the table means your organization will have a say in key decisions

For more information on joining the U.S. Technical Advisory Committee (TAG) for ISO TC 323 Circular Economy, please contact:
â€¢ US TAG ISO TC 323 chair Alison Conroy, Georgia Pacific (Alison.conroy@gapac.com) or US TAG ISO TC 323 chair-elect Mike Levy, First Environment (mlevy@firstenvironment.com)
â€¢ US TAG ISO TC 323 Secretary Emily Lorenz, Consulting Engineer (emilyblorenz@gmail.com)
â€¢ Travis Murdoch, ASTM International US TAG ISO TC 323 Administrator for US TAG (tmurdock@astm.org

The Sustainable Plastics Economy

Conference: Sustainable Packaging Symposium (SPS)

Year: 2019

Proceeding: Sustainable Packaging Symposium 2019

Group: Abstract Submissions

Session: Technical Session 3 Innovative Technologies

Time: Wednesday, December 18, 2019 - 10:45am-11:15am

Authors:

Clark, T., ENSO Plastics

The Circular Economy was created from the idea that we can create sustainable human systems by mimicking the sustainable processes in nature. A methodology to ensure resources flow continually between systems, where the waste from one process becomes a resource for another. The Sustainable Plastics Economy adopts this philosophy and creates a system without plastic waste that can ultimately integrate into nature, and most importantly can be implemented by organizations immediately.

In the plastic industry, the Circular Economy is often mistaken for a rebranding of the â €œrecycling agendaâ€. Recycling is typically understood as using a waste product to manufacture a similar product; such as using discarded water bottles to manufacture new water bottles. The primary focus of recycling is for plastic to remain in a continual cycle of use and recycling. This interpretation follows the traditional approach of increasing plastic recycle rates without addressing the technical limitations of the material. It also lacks the vision and robustness of the Circular Economy.

The Sustainable Plastics Economy integrates a complete Circular Economy approach with the unique challenges of plastic. It includes the concepts of Sustainable Materials Management by addressing the full LCA impact of various plastic options such as, what types of materials to select, where to source raw ingredients, waste infrastructures, and customary discard scenarios. The Sustainable Plastics Economy creates a dynamic, data driven approach to create a system designed to replicate and ultimately integrate into nature, as intended in the Circular Economy precept.

AUTHOR INDEX

AIChE
120 Wall Street, FL 23
New York, NY 10005-4020

CURRAN ASSOCIATES INC.
proceedings.com

ISBN 978-1-7138-0733-9

9 781713 807339